CANINE MEDICINE - RECENT TOPICS AND ADVANCED RESEARCH

Edited by **Hussein Abdelhay Elsayed Kaoud**

Canine Medicine - Recent Topics and Advanced Research

http://dx.doi.org/10.5772/63036
Edited by Hussein Abdelhay Elsayed Kaoud

Contributors

Ali Risvanli, Halis Ocal, Cahit Kalkan, Cleuza Maria de Faria Rezende, Changbaig Hyun, Rita Payan Carreira, Paulo Borges, Alain Fontbonne, Mitzi Sarahi Anaya Garcia, Jael Sarahi Hernandez Anaya, Chao-Nan Lin, Shu-Yun Chiang, Marcia Almeida Melo, Raizza Barros Sousa Silva, Laysa Freire Franco E Silva, Beatriz Maria De Almeida Braz, Jaqueline Maria Dos Santos Sousa, Saul José Semião Santos, Paulo Paes De Andrade, Akane Tanaka, João Antunes, Débora Alves De Carvalho Freire, Ilanna Vanessa Pristo De Medeiros Oliveira, Gabriela Hémylin Ferreira Moura, Larissa De Castro Demoner, Heider Irinaldo Pereira Ferreira

CBS, Edition **2018**

Published by InTech

Janeza Trdine 9, 51000 Rijeka, Croatia

Publishing Process Manager Maja Bozicevic

Technical Editor SPi Global

Cover Designer InTech Design Team

Additional hard copies can be obtained from orders@intechopen.com

Canine Medicine - Recent Topics and Advanced Research, Edited by Hussein Abdelhay Elsayed Kaoud
p. cm.
Print ISBN 978-953-51-2831-1
Online ISBN 978-953-51-2832-8

Contents

Preface

Dogs perform many roles for people, such as hunting, herding, pulling loads, protection, assisting police and military, companionship, and aiding handicapped individuals. This influence on human society has given them the sobriquet "man's best friend."

There is a possibility that during a pet's lifetime, medication may be recommended to treat medical conditions or problems. So, it does not hurt to know what the most common medical conditions or problems are and how much you might pay for treatment.

This book *Canine Medicine - Recent Topics and Advanced Research* provides the knowledge in diagnosis and treatment of some important diseases and problems that the canines face.

I believe that this book offers broader perspective to the readers in the recent advances in **canine medicine**, starting from recent topics to application in clinical diagnosis and therapeutics for practitioners and veterinarians.

The main purpose of the book is to point out the interest of some important topics of canine medicine and the progress in this field and to clear its importance in veterinary medicine. The book is concisely and clearly written and intended for veterinarians and clinicians and for people directly involved in dog health, management, and breeding.

The book included six sections ((1) Infectious Diseases, (2) Oncology, (3) Cardiology, (4) Reproduction, (5) Orthopedics, and (6) Diagnostic Neurology) and has nine chapters.

Chapter 1: "Canine Parvovirus Type 2 (CPV-2)." The authors outline the importance of the disease and how to prevent and control it. Enteritis is characterized by intestinal hemorrhage with severe bloody diarrhea. The causative agent, CPV-2, was first identified in the late 1970s. CPV is a nonenveloped, linear, single-stranded DNA virus with a genome of approximately 5 kb, and it belongs to the genus *Parvovirus* , together with feline panleukopenia virus, mink enteritis virus, raccoon parvovirus, and porcine parvovirus. An antigenic variant, CPV-2a, was identified within a few years after the emergence of CPV-2, and another variant, CPV-2b, began appearing in the canine population in 1984. In 2000, a novel antigenic variant, CPV-2c, was first detected in Italy. This chapter focuses on the history, viral evolution, epidemiology, pathogenesis, clinical signs, diagnosis, vaccination, and prevention of CPV-2.

Chapter 2: "Canine Visceral Leishmaniasis (CVL) in Brazil."—neglected diseases, risk factors, zoonoses." In this chapter, the authors describe the cause and its transmission through the bite of sand flies, as well as its diagnosis, therapy, and prevention. Endemic regions have widened, and canine visceral leishmaniasis (CVL) occurs mainly in the Mediterranean region and South America. There is no consensus on the risk factors associated with CVL, as results differ between the studied regions and countries. This chapter describes the main aspects of epidemiology, immunology, clinical signs, diagnosis treatment, and control of canine visceral leishmaniasis with emphasis on Brazil.

Chapter 3: "Infectious Causes of Abortion, Stillbirth, and Neonatal Death in Bitches."—puppies, bitches, reproductive failure. Problems in gestational development in dogs can be determined by infectious and noninfectious causes. Trauma during pregnancy, genetic characteristics of the animal, deficit nutrition, thyroid dysfunction, maternal problems, and hormonal disorders are among the noninfectious causes found. The majority of the cases are in relation to infectious causes; one should consider viral, bacterial, fungal, and protozoal, which can interfere directly or indirectly in fetal development. The progression of fetal development may be affected by the direct action of microorganisms to overcome the placenta, but they are also able to affect pregnancy and placental toxins by inflammatory processes and may still occur in maternal pathologies, which entail problems such as hyperthermia, hypoxia, and endotoxemia, which can result in abortion. Several diseases can trigger pregnancy loss in dogs. This action can be direct by microorganisms, as well as indirectly triggering other problems that lead to abortion. This chapter discusses the infectious etiologies of reproductive failures (abortion, stillbirth, and neonatal death) in bitches.

Chapter 4: "Mast Cell Tumors."—mast cells, tumorigenesis, pathology, treatment. This chapter describes one of the major cutaneous tumors in dogs. Though the etiology of MCTs are not completely understood, it becomes clear that approximately 10 to 20% MCTs express mutant KIT receptors with ligand-independent phosphorylation. Tyrosine kinase inhibitors targeting KIT exert antitumor effects on malignant proliferation of mast cells with or without gene mutations. However, the efficacy of KIT inhibitors on dogs with MCTs has been limited. In this chapter, the authors try to outline the general understandings of mast cells such as the process of its differentiation and proliferation and what has been revealed regarding the mechanism of tumorigenesis and therapeutic approaches. Especially, KIT mutation-related evidences and therapeutic approaches in the future are discussed.

Chapter 5: "Chronic Mitral Valve Insufficiency in Dogs: Recent Advances in Diagnosis and Treatment."—mitral valve, degenerative valve disease, mitral valve insufficiency, heart failure, mitral regurgitation. Chronic mitral valvular insufficiency (CMVI) is the most common acquired heart disease in dogs and is characterized by degenerative valvular changes causing progressive thickening of mitral leaflets and incomplete closure of mitral valve. As the disease progresses, it causes congestive heart failure and pulmonary edema if the LA dilation cannot accommodate the volume overload by mitral regurgitation. Therefore, it is the most common cause of cardiac mortality in dogs. This chapter discusses general features of CMVI in dogs focusing on recent advances in diagnosis and treatment.

Chapter 6: "Abnormalities in the Sexual Cycle of Bitches."—Sexual cycle abnormalities are an important cause of infertility in bitches, with disorders such as anestrous, split estrus, and persistent estrus having varied etiologies. Sexual cycle abnormalities in bitches may be addressed as follicular or luteal-phase disorders. However, pet owners should have a good working knowledge of the sexual cycles of their animals in order to better understand these disorders.

Chapter 7: "Pursuing Alternative Strategies for Healthier Medical Contraception in Dogs." —medical contraception, nonprogestogen drugs, antigonadotropins. Though extensively used in the control of the reproductive cycles in either the domestic or feral dogs, as well as in wild carnivores, medical progestin-based contraception still raises concerns to the veterinary practitioner and owners on its safety and efficiency. These concerns endorsed, in last decades, the research in the development of new alternatives for effective, reversible, and safe contraceptive methods for carnivores, mainly pursuing a larger scale control of canine reproduction and the development of products with few side effects. Nowadays, the medical contraceptives often intend to master, in a reversible way, the reproductive cycle in genetically valuable dogs, which presumes that they would be active for short periods of time and ought to safeguard the animal fertility. However, hormonal contraceptives are also used worldwide to control the reproductive activity in either domestic or feral cats, for long-term treatments, because of a pretended short-term economic interest. Progestogens are the most frequently used hormonal contraceptive in carnivores. They are rather easy to obtain across the globe and relatively cheap; they have diverse drug presentations, allowing its use independent of veterinary assistance, and are effective in preventing pregnancy. Still a significant number of undesirable health side effects are attributed to progestins when employed with some chronicity, when applied in older animals, or even when it is due to a misuse. In the past two decades, several new approaches to managing dog reproduction were proposed to escape progestins. However, their efficiency and cost are still to be proven as a viable alternative around the world. This chapter aims to review the medical methods available as alternative to the progestin in canine contraception, addressing particularly the future perspectives, opportunities, and limitations linked to currently available substitutes, based on our practice. This information can be of utmost interest to students, clinicians, or colonies' technicians.

Chapter 8: "Patellar Luxation in Small Animals."—patella, surgery, radiography, dog, cat. This chapter describes lesions that occur in the stifle joints of dogs with patellar luxation. These lesions are associated with the animal's age, body weight, and degree of luxation. It also reports on the rate of redislocation. The patellar lesions found include articular cartilage erosion, subchondral bone exposure, a flattened or concave patellar surface, and enthesophytes. Extrapatellar lesions included synovitis, osteophytes, blunting of the trochlear groove, an absence of trochlea, erosion of the condylar margins, capsule thickening, a long digital extensor tendon injury, "joint mice," flap formation, cranial cruciate ligament rupture, and meniscal prolapse. Such lesions were frequently found in animals with Grade II or III luxation that were aged 24 months or older; they were more severe in dogs weighing more than 15 kg. Patellar luxation causes changes that favor articular degeneration and should be treated surgically. Conservative treatment relieves pain but does not address tissue alterations.

Chapter 9: "Diffusion Tensor Tractography in Cerebral White Matter."—diffusion tensor, cerebral white matter, in vivo. Conventional magnetic resonance imaging (MRI) allows researchers and doctors to observe the autopsy and injuries of the cerebral white matter (CWM) in dogs. However, there are dynamic images on the basis of need (DT) tensor deployment of fiber to assess the safety of the device from the CWM technology. Diffusion tensor tractography (DTT) produces a three-dimensional representation of the data, which is displayed on a color map obtained from the variation of water molecules in space CWM. Fractional anisotropy (FA) variation is a value that measures changes in the deployment of water, which can occur if spaces CWM displaced disrupted, or infiltration. The aim of this chapter is to determine the feasibility of DTT in vivo examination of a natural appearance of the CWM in dogs by visual and quantitative analysis of the most representative areas of CWM.

The book is the result of the collaborating parties. I gratefully acknowledge the assistance provided by all authors that have contributed to the publication of this volume and by the InTech editorial office that initiated this project and that has completed the book edition.

Dr. Hussein Abdelhay Elsayed Kaoud
Full Professor of Animal, Poultry, Aquaculture Health, and Environmental Pollution
Department of Hygiene and Management
Faculty of Veterinary Medicine
Cairo University
El-Giza, Egypt

Section 1

Infectious Diseases

Chapter 1

Canine Parvovirus Type 2

Chao-Nan Lin and Shu-Yun Chiang

Additional information is available at the end of the chapter

http://dx.doi.org/10.5772/65801

Abstract

Canine parvovirus (CPV) enteritis is characterized by intestinal hemorrhage with severe bloody diarrhea. The causative agent, CPV-2, was first identified in the late 1970s. CPV is a nonenveloped, linear, single-stranded DNA virus with a genome of approximately 5 kb, and it belongs to the genus *Parvovirus*, together with feline panleukopenia virus, mink enteritis virus, raccoon parvovirus, and porcine parvovirus. An antigenic variant, CPV-2a, identified within a few years after the emergence of CPV-2, and another variant, CPV-2b, began appearing in the canine population in 1984. In 2000, a novel antigenic variant, CPV-2c, was first detected in Italy. This chapter focuses on the history, viral evolution, epidemiology, pathogenesis, clinical signs, diagnosis, vaccination, and prevention of CPV-2.

Keywords: canine parvovirus type 2, CPV-2, viral evolution, antigenic variants, epidemiology, pathogenesis

1. Introduction

Canine parvovirus (CPV) infection is characterized by clinical gastroenteritis with severe hemorrhagic diarrhea and is a common infectious disease in younger dogs [1]. Gastroenteritis caused by CPV type 2 (CPV-2) infection, especially that caused by the newer variants of CPV-2, may progress rapidly [2, 3]. Dehydration is the rapid onset of the loss of bodily fluids caused by severe vomiting, diarrhea, or hemorrhagic diarrhea. Death occurs in as early as a few days after disease onset. This chapter documents the history, viral evolution, epidemiology, pathogenesis, clinical signs, diagnosis, vaccination, and prevention of CPV-2.

1.1. History

Canine parvoviral enteritis is characterized by intestinal hemorrhage with severe bloody diarrhea [4]. The causative agent, CPV-2, was first identified in the late 1970s [5]. CPV is a nonenveloped, linear, single-stranded DNA virus with a genome of approximately 5 kb, and it belongs to the genus *Parvovirus,* together with feline panleukopenia virus (FPV), mink enteritis virus, raccoon parvovirus, and porcine parvovirus [6]. CPV-2 is distinct from CPV-1, also known as canine minute virus, which belongs to the genus *Bocavirus.* Indeed, CPV-2 is believed to have originated from FPV or a closely related FPV-like parvovirus of wild carnivores (**Figure 1**) [7, 8]. Various hypotheses for how this may have occurred have been suggested, including direct mutation from FPV and contact between cats and dogs kept as companion animals within the same home [8].

1.2. Viral evolution

Shortly after CPV was first identified in 1978, the original virus, CPV-2, was subsequently replaced in the dog population by strains carrying small antigenic variations (termed 2a, 2b, and 2c) of the VP2 protein that could be distinguished by monoclonal antibodies and genetic analysis. An antigenic variant, CPV-2a, identified within a few years after the emergence of CPV-2 [9, 10], and another variant, CPV-2b, began appearing in the canine population in 1984 [11]. A novel antigenic variant, Glu-426 mutant (now termed CPV-2c), was first detected in Italy in 2000 [12]. New antigenic variants of CPV-2 have been observed in epidemics around the world and are soon replacing the original CPV-2. CPV-2a variant shows several substitutions within the VP2 protein, including Met87Leu, Ile101Thr, Ala300Gly, and Asp305Tyr. Furthermore, CPV-2b variant has been identified to contain an additional amino acid change, Asn426Asp [13, 14]. These two antigenic variants further evolved into new CPV-2a and -2b

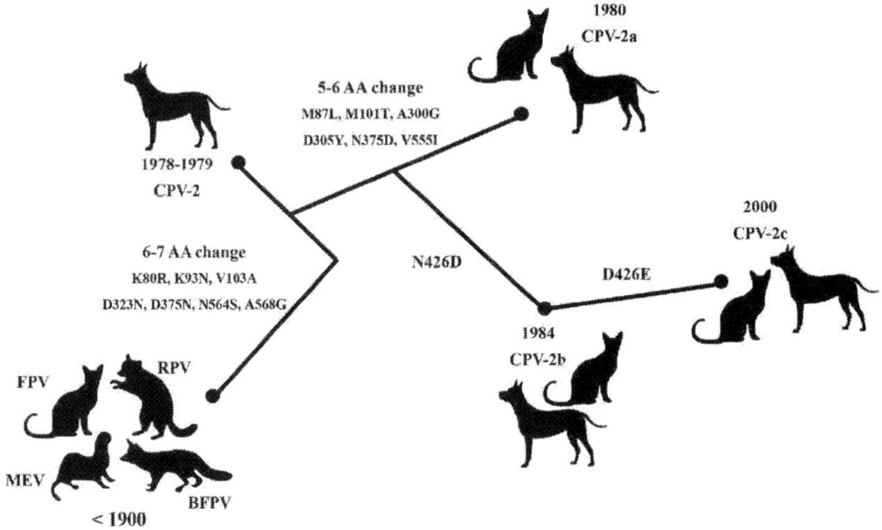

Figure 1. Cartoon of the evolution of canine parvovirus type (CPV-2). CPV-2 is believed to have originated from FPV or a closely related FPV-like parvovirus of wild carnivores (modified from Truyen [8]).

types, with amino acid substitutions at residue 297 (Ser to Ala), during the 1990s [15]. Antigenic variant CPV-2c was identified only one residue substitution (Asp426Glu) as differ from CPV-2a and -2b [12].

In addition to the amino acid substitution described above, amino acid substitutions Tyr324Ile, Gln370Arg, and Thr440Ala in the surface of the VP2 were also noted in the recent year [16]. Review of CPV-2 sequences from GenBank showed that this Ile324 variant of CPV-2a is also found in several countries, including Korea [17, 18], China [19–25], Thailand [26], Uruguay [27, 28], Japan [29], Taiwan [30, 31], and India [32, 33]. Surprisingly, aside from the Uruguayan strains, this Ile324 CPV-2a variant is only distributed in Asian countries. Previous study has shown that residue 324 of VP2 is subject to positive selection among all carnivore parvoviruses [34]. This residue is adjacent to the residue 323, which together with residue 93 is known to be affected in host range and tropism of canine transferrin receptor binding [35]. The mutation of VP2 residue 323 may affect interactions between residues in neighboring loops of either the same VP2 molecule or the threefold-related VP2, greatly reducing replication in canine cells [36]. The Gln370Arg change is unique in the Taiwanese [85] and Chinese CPV-2c strains [22, 25]. This substitution is also observed in Chinese panda parvovirus [37]. Residue 370 is located between residues 359 and 375, which constitutes a flexible surface loop of the capsid protein that is adjacent to a double Ca^{2+}-binding site. They were found to be essential for virus infectivity. Changes in them are correlated with the ability of the virus to hemagglutinate erythrocytes [38]. In addition, VP2 position 440 is located near a major antigenic site. The Thr440Ala change was found in China, the USA [39], Italy [14], Argentina [40], Uruguay [27], India [33], and Taiwan [85].

1.3. Epidemiology

According to the epidemiology surveillance of CPV-2, the distribution of these three antigenic variants (-2a, -2b, and -2c) is summarized in **Table 1**. Previous study has revealed that the oldest CPV-2c variant was isolated in 1996 in Germany [41]. Epidemiological surveillance in

Country	Positive of strain detected			Reference
	CPV-2a	CPV-2b	CPV-2c	
Europe				
Italy	+	+	+	[2]
Portugal	+	+	+	[16]
Spain	+	+	+	[78]
France	+	+	+	[79]
UK	+	+	+	[41]
Belgium	+	-	+	[79]
Germany	+	+	+	[79]
Greece	+	+	+	[43]

Country	Positive of strain detected			Reference
	CPV-2a	CPV-2b	CPV-2c	
Switzerland	+	+	+	[41]
Czech public	+	+	-	[79]
Romania	+	-	-	[79]
Hungary	+	-	-	[80]
Bulgaria	+	+	+	[81]
Turkey	+	+	+	[82]
Africa				
Tunisia	+	+	+	[47]
Morocco	+	+	+	[53]
North America				
USA	+	+	+	[39]
South America				
Uruguay	+	+	+	[83]
Argentina	+	+	+	[50]
Brazil	+	+	+	[84]
Ecuador	+	+	+	[51]
Mexico	-	-	+	[52]
Asia				
India	+	+	+	[55]
Taiwan	+	+	+	[85]
Korea	+	+	-	[18]
Japan	+	+	-	[15]
China	+	+	+	[19]
Thailand	+	+	-	[26]
Vietnam	+	+	+	[54]
Oceania		+	+	-
Australia	+	+	-	[86]

Table 1. Detection of the canine parvovirus variants in the world.

Europe shows that CPV-2c is predominant currently in Italy, Germany, and Spain and is also extensively co-distributed with CPV-2a or -2b in Portugal [42], Belgium, France, Greece [43], Bulgaria [44], Sweden [45], Turkey [46], and the United Kingdom. In recent years, CPV-2c has also been widespread in Tunisia [47], the USA [39], Uruguay [48], Brazil [49], Argentina [50],

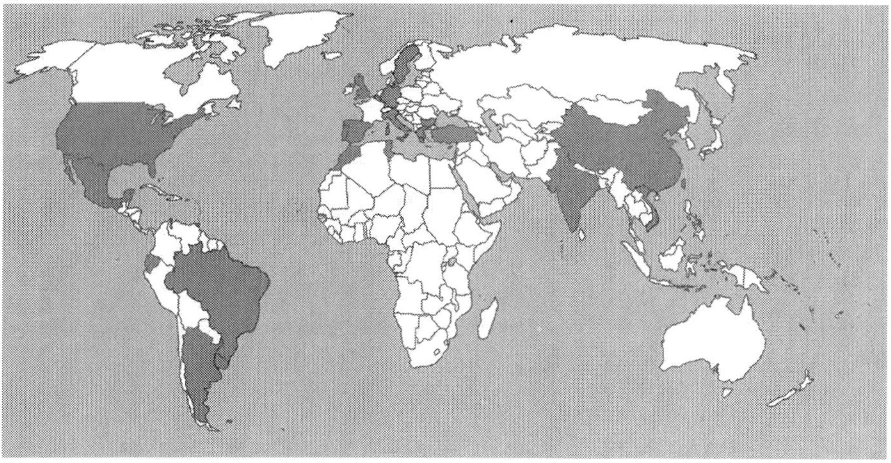

Figure 2. Distribution of canine parvovirus type 2c (CPV-2c) variants around the world. Red color represents that countries reporting CPV-2c case.

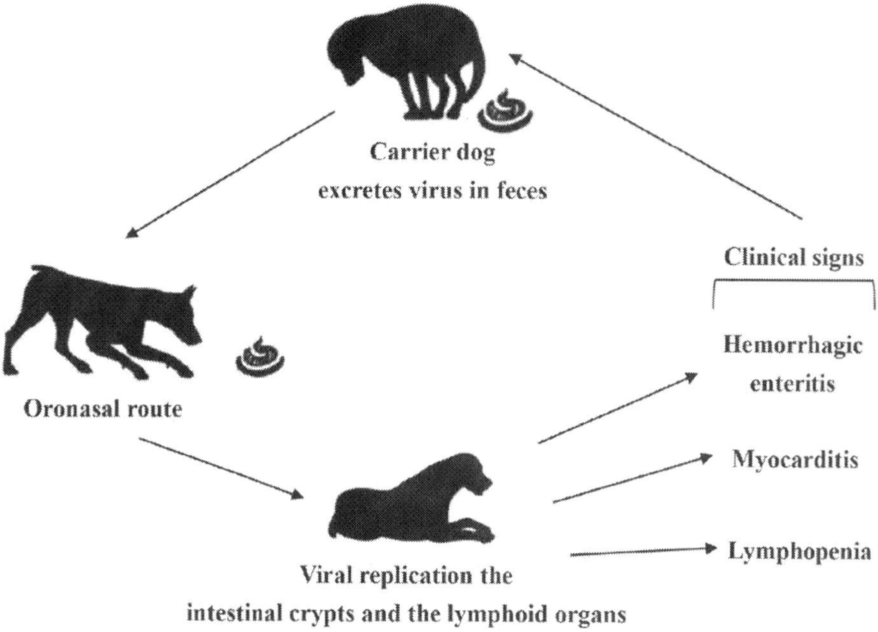

Figure 3. Transmission and clinical symptoms of canine parvovirus type 2 (CPV-2). Transmission of CPV-2 occurred by the fecal-oral route, after exposure to CPV-2 in feces and/or vomit, or viral-contaminated fomites.

Ecuador [51], Mexico [52], and Morocco [53]. Surprisingly, the CPV-2c variant has not been prevalent in Asia since the first identified in Vietnam in 2004 [54]. Only a few CPV-2c strains have been reported in India [55], China [19, 22, 25], and Taiwan [85]. To date, either CPV-2a or -2b has been prevalent in Asian countries [15, 17–19, 22, 25, 26, 29–33, 55–59]. Phylogenetic analysis demonstrated that the recent CPV-2c isolate from Taiwan shares a common evolutionary origin with Chinese strains of CPV-2c, as classified into novel Asian CPV-2c variants (Phe267Tyr, Tyr324Ile, Gln370Arg, and Asp426Glu) [85]. **Figure 2** summarizes the distribution of the CPV-2c variant around the world.

1.4. Pathogenesis

Transmission of CPV-2 occurred by the fecal-oral route, after exposure to CPV-2 in feces and/or vomit, or viral-contaminated fomites (**Figure 3**). The severity of clinical symptoms depends on the factors such as viral strain, host immunity, and the presence of the coinfection with other pathogens. In naturally [60] or experimentally [61] infected dogs, many dogs never develop overt clinical symptoms, especially when the dogs have high level of the maternally derived antibodies (MDAs) [61]. Higher titers of MDA in young dogs are protective against infection by CPV-2. Infected dogs with severe clinical symptoms of disease had lower levels of hemagglutination inhibition (HI) titers than did animals without clinical signs following

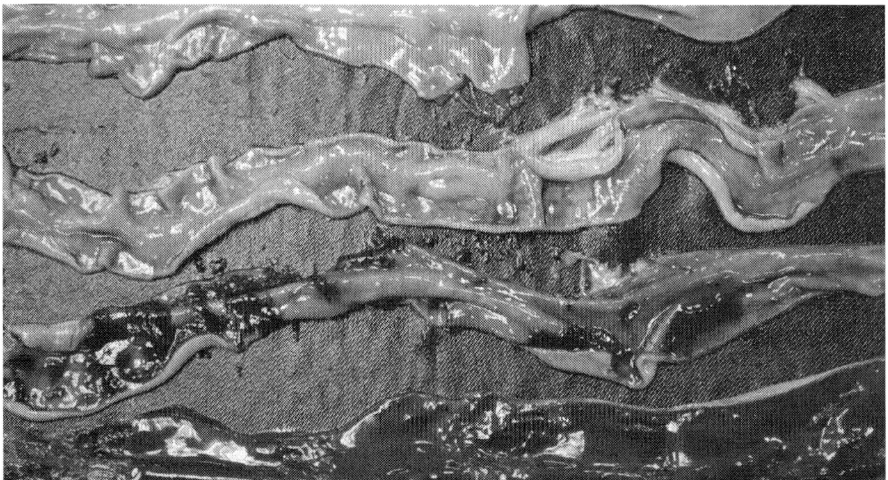

Figure 4. Mucosal hemorrhage in canine parvovirus type 2 (CPV-2)-infected dog. A 4-month-old puppy suffered from CPV-2 which showed lethargy, inappetence, and bloody diarrhea (from Professor Ming-Tang Chiou, Department of Veterinary Medicine, National Pingtung University of Science and Technology, Taiwan).

CPV-2 infection [61]. In the lowest HI titers group, CPV-2 shedding was noted up to 45 days post infection [61]. The incubation period of CPV-2 infection is 1–2 weeks. The affected gastrointestinal tissues include epithelium of the tongue, oral cavity, esophagus, and intestinal tract. CPV-2 replicates and destroys epithelial cells of the intestinal crypts causing malabsorption and increased intestinal permeability. CPV-2 shedding in feces was detected up to 45 and 54 days in experimentally [61] and naturally infected dogs [60], respectively. The period could also potentially correlate with the severity of the clinical status of the animals [62]. Viral shedding from infected dogs with bloody diarrhea was observed up to 63 days compared to 54 days in infected dogs with diarrhea (no dogs with hemorrhagic diarrhea or vomiting were tested) [60].

1.5. Clinical signs

The clinical signs of CPV-2 infection include fever, lethargy, inappetence, vomiting, diarrhea or bloody diarrhea (**Figure 4**), and dehydration. Myocarditis usually occurs up to the first 4 weeks of age or infected in utero, which may result in signs of sudden death

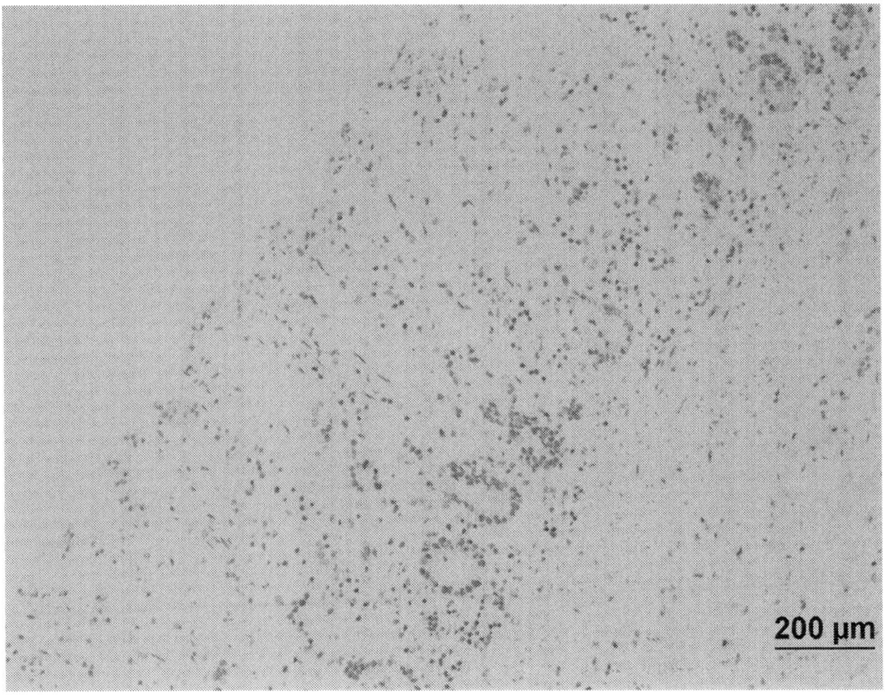

Figure 5. Detection of canine parvovirus type 2 (CPV-2) antigens in intestinal tract using in situ hybridization. CPV-2 immunolabeling is seen in the crypt epithelium (from Professor Ming-Tang Chiou, Department of Veterinary Medicine, National Pingtung University of Science and Technology, Taiwan).

or congestive heart failure. Cerebellar hypoplasia has been rarely reported in dogs with utero infection. CPV-2c shows almost the same clinical signs as CPV-2a and CPV-2b, such as anorexia, vomiting, acute gastroenteritis, and hemorrhagic diarrhea. However, CPV-2c infection has been reported to be indicative of a more severe disease induced by this variant [51, 63].

1.6. Diagnosis

The most common abnormalities found on the CBC are leukopenia, lymphopenia, and neutropenia. Some disease dogs develop anemia as a result of gastrointestinal blood loss. Electrolyte and coagulation abnormalities have been reported in dogs with parvoviral enteritis. The diagnosis of CPV-2 infection has relied on probe-based real-time polymerase chain reaction (PCR) [64–69], SYBR green-based real-time PCR [62, 70–72], conventional PCR [47, 68], electron microscopy [73], and methods provided as commercial kits [74]. CPV–2 antigens were found in crypt epithelium (**Figure 5**) or affected tissues such as spleen (**Figure 6**) using in situ hybridization.

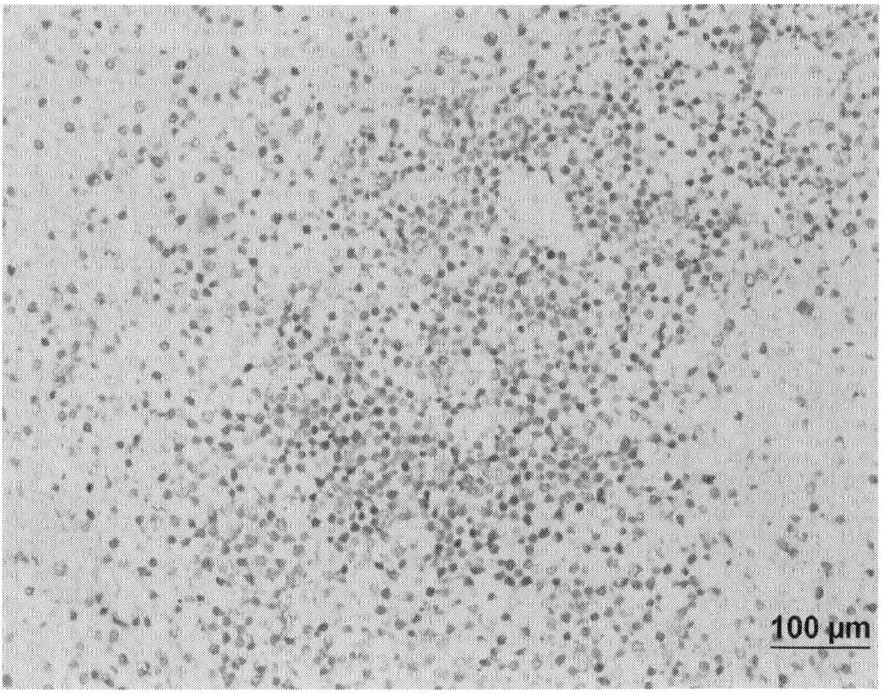

Figure 6. Detection of canine parvovirus type 2 (CPV-2) antigens in spleen using in situ hybridization. CPV-2 immunolabeling is seen in lymphocyte (from Professor Ming-Tang Chiou, Department of Veterinary Medicine, National Pingtung University of Science and Technology, Taiwan).

1.7. Vaccination and prevention

Immunization is the most effective method for the prevention of CPV-2 enteritis. However, the initial immunization should be of concern regarding the interference of the MDA. The second concern is whether the currently available CPV-2 vaccine provides adequate protection against CPV-2c infection. VP2 encodes a viral capsid protein that is the major structural protein of CPV-2 and is involved in the host-immune response [75]. Therefore, a small number of mutations may result in increased pathogenicity [51]. Several studies have demonstrated the efficacy of the current CPV-2 vaccine against CPV-2c infection [76, 77]. By contrast, some evidence suggests that dogs with the complete vaccination program still suffer from CPV-2c [63]. Therefore, the efficacy of the current vaccine against prototype CPV-2c and/or novel CPV-2c variant remains to be evaluated, especially in regard to the amino acid substitutions observed in the novel CPV-2c variant as compared to the prototype of CPV-2c. Parvoviruses are extremely stable in the environment and can be transmitted via indirect contact, an important factor in their maintenance in populations. Several disinfectants had been reported that parvoviruses can be inactivated with a 1:30 dilution of household bleach, potassium peroxymonosulfate, and accelerated hydrogen peroxide. These disinfectants will also inactivate other viruses.

2. Conclusion

CPV-2 is most likely one of the most common infectious diseases in younger dogs. CPV-2 is constantly mutating, leading to the evolution of novel variants. All countries should continue surveillance and monitor the events associated with CPV-2 disease. The need to update current vaccines remains to be assessed.

Author details

Chao-Nan Lin* and Shu-Yun Chiang

*Address all correspondence to: cnlin6@mail.npust.edu.tw

Department of Veterinary Medicine, College of Veterinary Medicine, National Pingtung University of Science and Technology, Pingtung, Taiwan, ROC

References

[1] Yesilbag K, Yilmaz Z, Ozkul A, Pratelli A: Aetiological role of viruses in puppies with diarrhoea. Vet Res 2007, **161**:169–170.

[2] Buonavoglia C, Martella V, Pratelli A, Tempesta M, Cavalli A, Buonavoglia D, Bozzo G, Elia G, Decaro N, Carmichael L: Evidence for evolution of canine parvovirus type 2 in Italy. J Gen Virol 2001, **82**:3021–3025.

[3] Martella V, Cavalli A, Pratelli A, Bozzo G, Camero M, Buonavoglia D, Narcisi D, Tempesta M, Buonavoglia C: A canine parvovirus mutant is spreading in Italy. J Clin Microbiol 2004, **42**:1333–1336.

[4] MacLachlan NJ, Dubovi EJ: Parvoviridae. In *Fenner's Veterinary Virology*. 4th edn. Edited by MacLachlan NJ, Dubovi EJ. London: Academic Press; 2011: 507.

[5] Appel MJ, Scott FW, Carmichael LE: Isolation and immunisation studies of a canine parco-like virus from dogs with haemorrhagic enteritis. Vet Res 1979, **105**:156–159.

[6] Hoelzer K, Parrish CR: The emergence of parvoviruses of carnivores. Vet Res 2010, **41**:39–51.

[7] Allison AB, Harbison CE, Pagan I, Stucker KM, Kaelber JT, Brown JD, Ruder MG, Keel MK, Dubovi EJ, Holmes EC, Parrish CR: Role of multiple hosts in the cross-species transmission and emergence of a pandemic parvovirus. J Virol 2012, **86**:865–872.

[8] Truyen U: Evolution of canine parvovirus-a need for new vaccines? Vet Microbiol 2006, **117**:9–13.

[9] Parrish CR, O'Connell PH, Evermann JF, Carmichael LE: Natural variation of canine parvovirus. Science 1985, **230**:1046–1048.

[10] Parrish CR, Have P, Foreyt WJ, Evermann JF, Senda M, Carmichael LE: The global spread and replacement of canine parvovirus strains. J Gen Virol 1988, **69(Pt 5)**: 1111–1116.

[11] Parrish CR, Aquadro CF, Strassheim ML, Evermann JF, Sgro JY, Mohammed HO: Rapid antigenic-type replacement and DNA sequence evolution of canine parvovirus. J Virol 1991, **65**:6544–6552.

[12] Buonavoglia C, Martella V, Pratelli A, Tempesta M, Cavalli A, Buonavoglia D, Bozzo G, Elia G, Decaro N, Carmichael L: Evidence for evolution of canine parvovirus type 2 in Italy. J Gen Virol 2001, **82**:3021–3025.

[13] Martella V, Decaro N, Buonavoglia C: Evolution of CPV-2 and implication for antigenic/genetic characterization. Virus Genes 2006, **33**:11–13.

[14] Decaro N, Desario C, Parisi A, Martella V, Lorusso A, Miccolupo A, Mari V, Colaianni ML, Cavalli A, Di Trani L, Buonavoglia C: Genetic analysis of canine parvovirus type 2c. Virology 2009, **385**:5–10.

[15] Ohshima T, Hisaka M, Kawakami K, Kishi M, Tohya Y, Mochizuki M: Chronological analysis of canine parvovirus type 2 isolates in Japan. J Vet Med Sci 2008, **70**:769–775.

[16] Miranda C, Thompson G: Canine parvovirus: the worldwide occurrence of antigenic variants. J Gen Virol 2016, **97**: 2043-2057..

[17] Jeoung SY, Ahn SJ, Kim D: Genetic analysis of VP2 gene of canine parvovirus isolates in Korea. J Vet Med Sci 2008, **70**:719–722.

[18] Yoon SH, Jeong W, Kim HJ, An DJ: Molecular insights into the phylogeny of canine parvovirus 2 (CPV-2) with emphasis on Korean isolates: a Bayesian approach. Arch Virol 2009, **154**:1353–1360.

[19] Zhang R, Yang S, Zhang W, Zhang T, Xie Z, Feng H, Wang S, Xia X: Phylogenetic analysis of the VP2 gene of canine parvoviruses circulating in China. Virus Genes 2010, **40**:397–402.

[20] Yi L, Tong M, Cheng Y, Song W, Cheng S: Phylogenetic analysis of canine parvovirus VP2 gene in China. Transbound Emerg Dis 2016, 63:e262-e269.

[21] Zhong Z, Liang L, Zhao J, Xu X, Cao X, Liu X, Zhou Z, Ren Z, Shen L, Geng Y, et al: First isolation of new canine parvovirus 2a from Tibetan mastiff and global analysis of the full-length VP2 gene of canine parvoviruses 2 in China. Int J Mol Sci 2014, **15**:12166–12187.

[22] Geng Y, Guo D, Li C, Wang E, Wei S, Wang Z, Yao S, Zhao X, Su M, Wang X, et al: Co-circulation of the rare CPV-2c with unique Gln370Arg substitution, new CPV-2b with unique Thr440Ala substitution, and new CPV-2a with high prevalence and variation in Heilongjiang Province, Northeast China. PLoS One 2015, **10**:e0137288.

[23] Han SC, Guo HC, Sun SQ, Shu L, Wei YQ, Sun DH, Cao SZ, Peng GN, Liu XT: Full-length genomic characterizations of two canine parvoviruses prevalent in Northwest China. Arch Microbiol 2015, **197**:621–626.

[24] Xu J, Guo HC, Wei YQ, Shu L, Wang J, Li JS, Cao SZ, Sun SQ: Phylogenetic analysis of canine parvovirus isolates from Sichuan and Gansu provinces of China in 2011. Transbound Emerg Dis 2015, **62**:91–95.

[25] Zhao H, Wang J, Jiang Y, Cheng Y, Lin P, Zhu H, Han G, Yi L, Zhang S, Guo L, Cheng S: Typing of canine parvovirus strains circulating in North-East China. Transbound Emerg Dis 2015. doi:10.1111/tbed.12390.

[26] Phromnoi S, Sirinarumitr K, Sirinarumitr T: Sequence analysis of VP2 gene of canine parvovirus isolates in Thailand. Virus Genes 2010, **41**:23–29.

[27] Perez R, Bianchi P, Calleros L, Francia L, Hernandez M, Maya L, Panzera Y, Sosa K, Zoller S: Recent spreading of a divergent canine parvovirus type 2a (CPV-2a) strain in a CPV-2c homogenous population. Vet Microbiol 2012, **155**:214–219.

[28] Perez R, Calleros L, Marandino A, Sarute N, Iraola G, Grecco S, Blanc H, Vignuzzi M, Isakov O, Shomron N, et al: Phylogenetic and genome-wide deep-sequencing analyses of canine parvovirus reveal co-infection with field variants and emergence of a recent recombinant strain. PLoS One 2014, **9**:e111779.

[29] Soma T, Taharaguchi S, Ohinata T, Ishii H, Hara M: Analysis of the VP2 protein gene of canine parvovirus strains from affected dogs in Japan. Res Vet Sci 2013, **94**:368–371.

[30] Chou SJ, Lin HT, Wu JT, Yang WC, Chan KW: Genotyping of canine parvovirus type 2 VP2 gene in southern Taiwan in 2011. Taiwan Vet J 2013, **39**:81–92.

[31] Lin CN, Chien CH, Chiou MT, Chueh LL, Hung MY, Hsu HS: Genetic characterization of type 2a canine parvoviruses from Taiwan reveals the emergence of an Ile324 mutation in VP2. Virol J 2014, **11**:39.

[32] Mukhopadhyay HK, Matta SL, Amsaveni S, Antony PX, Thanislass J, Pillai RM: Phylogenetic analysis of canine parvovirus partial VP2 gene in India. Virus Genes 2013, **48**:89–95.

[33] Mittal M, Chakravarti S, Mohapatra JK, Chug PK, Dubey R, Narwal PS, Kumar A, Churamani CP, Kanwar NS: Molecular typing of canine parvovirus strains circulating from 2008-2012 in an organized kennel in India reveals the possibility of vaccination failure. Infect Genet Evol 2014, **23**:1–6.

[34] Hoelzer K, Shackelton LA, Parrish CR, Holmes EC: Phylogenetic analysis reveals the emergence, evolution and dispersal of carnivore parvoviruses. J Gen Virol 2008, **89**:2280–2289.

[35] Hueffer K, Parrish CR: Parvovirus host range, cell tropism and evolution. Curr Opin Microbiol 2003, **6**:392–398.

[36] Chang SF, Sgro JY, Parrish CR: Multiple amino acids in the capsid structure of canine parvovirus coordinately determine the canine host range and specific antigenic and hemagglutination properties. J Virol 1992, **66**:6858–6867.

[37] Guo L, Yang SL, Chen SJ, Zhang Z, Wang C, Hou R, Ren Y, Wen X, Cao S, Guo W, et al: Identification of canine parvovirus with the Q370R point mutation in the VP2 gene from a giant panda (*Ailuropoda melanoleuca*). Virol J 2013, **10**:163.

[38] Simpson AA, Chandrasekar V, Hebert B, Sullivan GM, Rossmann MG, Parrish CR: Host range and variability of calcium binding by surface loops in the capsids of canine and feline parvoviruses. J Mol Biol 2000, **300**:597–610.

[39] Hong C, Decaro N, Desario C, Tanner P, Pardo MC, Sanchez S, Buonavoglia C, Saliki JT: Occurrence of canine parvovirus type 2c in the United States. J Vet Diagn Invest 2007, **19**:535–539.

[40] Calderon MG, Romanutti C, Wilda M, D'Antuono A, Keller L, Giacomodonato MN, Mattion N, La Torre J: Resurgence of canine parvovirus 2a strain in the domestic dog population from Argentina. J Virol Methods 2015, **222**:145–149.

[41] Decaro N, Desario C, Addie DD, Martella V, Vieira MJ, Elia G, Zicola A, Davis C, Thompson G, Thiry E, et al: The study molecular epidemiology of canine parvovirus, Europe. Emerg Infect Dis 2007, **13**:1222–1224.

[42] Miranda C, Parrish CR, Thompson G: Epidemiological evolution of canine parvovirus in the Portuguese domestic dog population. Vet Microbiol 2016, **183**:37–42.

[43] Ntafis V, Xylouri E, Kalli I, Desario C, Mari V, Decaro N, Buonavoglia C: Characterization of Canine parvovirus 2 variants circulating in Greece. J Vet Diagn Invest 2010, **22**:737–740.

[44] Filipov C, Desario C, Patouchas O, Eftimov P, Gruichev G, Manov V, Filipov G, Buonavoglia C, Decaro N: A ten-year molecular survey on parvoviruses infecting carnivores in Bulgaria. Transbound Emerg Dis 2016, 63:460–464.

[45] Sutton D, Vinberg C, Gustafsson A, Pearce J, Greenwood N: Canine parvovirus type 2c identified from an outbreak of severe gastroenteritis in a litter in Sweden. Acta Vet Scand 2013, **55**:64.

[46] Muz D, Oguzoglu TC, Timurkan MO, Akin H: Characterization of the partial VP2 gene region of canine parvoviruses in domestic cats from Turkey. Virus Genes 2012, **44**:301–308.

[47] Touihri L, Bouzid I, Daoud R, Desario C, El Goulli AF, Decaro N, Ghorbel A, Buonavoglia C, Bahloul C: Molecular characterization of canine parvovirus-2 variants circulating in Tunisia. Virus Genes 2009, **38**:249–258.

[48] Perez R, Francia L, Romero V, Maya L, Lopez I, Hernandez M: First detection of canine parvovirus type 2c in South America. Vet Microbiol 2007, **124**:147–152.

[49] Pinto LD, Streck AF, Goncalves KR, Souza CK, Corbellini AO, Corbellini LG, Canal CW: Typing of canine parvovirus strains circulating in Brazil between 2008 and 2010. Virus Res 2012, **165**:29–33.

[50] Calderon MG, Romanutti C, D'Antuono A, Keller L, Mattion N, La Torre J: Evolution of canine parvovirus in Argentina between years 2003 and 2010: CPV2c has become the predominant variant affecting the domestic dog population. Virus Res 2011, **157**:106–110.

[51] Aldaz J, Garcia-Diaz J, Calleros L, Sosa K, Iraola G, Marandino A, Hernandez M, Panzera Y, Perez R: High local genetic diversity of canine parvovirus from Ecuador. Vet Microbiol 2013, **166**:214–219.

[52] Pedroza-Roldan C, Paez-Magallan V, Charles-Nino C, Elizondo-Quiroga D, De Cervantes-Mireles RL, Lopez-Amezcua MA: Genotyping of canine parvovirus in western Mexico. J Vet Diagn Invest 2015, **27**:107–111.

[53] Amrani N, Desario C, Kadiri A, Cavalli A, Berrada J, Zro K, Sebbar G, Colaianni ML, Parisi A, Elia G, et al: Molecular epidemiology of canine parvovirus in Morocco. Infect Genet Evol 2016, **41**:201–206.

[54] Nakamura M, Tohya Y, Miyazawa T, Mochizuki M, Phung HT, Nguyen NH, Huynh LM, Nguyen LT, Nguyen PN, Nguyen PV, et al: A novel antigenic variant of canine parvovirus from a Vietnamese dog. Arch Virol 2004, **149**:2261–2269.

[55] Nandi S, Chidri S, Kumar M, Chauhan RS: Occurrence of canine parvovirus type 2c in the dogs with haemorrhagic enteritis in India. Res Vet Sci 2010, **88**:169–171.

[56] Chang WL, Chang AC, Pan MJ: Antigenic types of canine parvoviruses prevailing in Taiwan. Vet Rec 1996, **138**:447.

[57] Wang HC, Chen WD, Lin SL, Chan JP, Wong ML: Phylogenetic analysis of canine parvovirus VP2 gene in Taiwan. Virus Genes 2005, **31**:171–174.

[58] Kang BK, Song DS, Lee CS, Jung KI, Park SJ, Kim EM, Park BK: Prevalence and genetic characterization of canine parvoviruses in Korea. Virus Genes 2008, **36**:127–133.

[59] Mohan Raj J, Mukhopadhyay HK, Thanislass J, Antony PX, Pillai RM: Isolation, molecular characterization and phylogenetic analysis of canine parvovirus. Infect Genet Evol 2010, **10**:1237–1241.

[60] Decaro N, Desario C, Campolo M, Elia G, Martella V, Ricci D, Lorusso E, Buonavoglia C: Clinical and virological findings in pups naturally infected by canine parvovirus type 2 Glu-426 mutant. J Vet Diagn Invest 2005, **17**:133–138.

[61] Decaro N, Campolo M, Desario C, Elia G, Martella V, Lorusso E, Buonavoglia C: Maternally-derived antibodies in pups and protection from canine parvovirus infection. Biologicals 2005, **33**:261–267.

[62] Lin CN, Chien CH, Chiou MT, Wang JW, Lin YL, Xu YM: Development of SYBR green-based real-time PCR for the detection of canine, feline and porcine Parvoviruses. Taiwan Vet J 2014, **40**:1–9.

[63] Decaro N, Desario C, Elia G, Martella V, Mari V, Lavazza A, Nardi M, Buonavoglia C: Evidence for immunisation failure in vaccinated adult dogs infected with canine parvovirus type 2c. New Microbiol 2008, **31**:125–130.

[64] Decaro N, Elia G, Martella V, Desario C, Campolo M, Trani LD, Tarsitano E, Tempesta M, Buonavoglia C: A real-time PCR assay for rapid detection and quantitation of canine parvovirus type 2 in the feces of dogs. Vet Microbiol 2005, **105**:19–28.

[65] Decaro N, Elia G, Desario C, Roperto S, Martella V, Campolo M, Lorusso A, Cavalli A, Buonavoglia C: A minor groove binder probe real-time PCR assay for discrimination between type 2-based vaccines and field strains of canine parvovirus. J Virol Methods 2006, **136**:65–70.

[66] Decaro N, Elia G, Martella V, Campolo M, Desario C, Camero M, Cirone F, Lorusso E, Lucente MS, Narcisi D, et al: Characterisation of the canine parvovirus type 2 variants using minor groove binder probe technology. J Virol Methods 2006, **133**:92–99.

[67] McKnight CA, Maes RK, Wise AG, Kiupel M: Evaluation of tongue as a complementary sample for the diagnosis of parvoviral infection in dogs and cats. J Vet Diagn Invest 2007, **19**:409–413.

[68] Mochizuki M, San Gabriel MC, Nakatani H, Yoshida M, Harasawa R: Comparison of polymerase chain reaction with virus isolation and haemagglutination assays for the detection of canine parvoviruses in faecal specimens. Res Vet Sci 1993, **55**:60–63.

[69] Chen HY, Li XK, Cui BA, Wei ZY, Li XS, Wang YB, Zhao L, Wang ZY: A TaqMan-based real-time polymerase chain reaction for the detection of porcine parvovirus. J Virol Methods 2009, **156**:84–88.

[70] Kumar M, Nandi S: Development of a SYBR Green based real-time PCR assay for detection and quantitation of canine parvovirus in faecal samples. J Virol Methods 2010, **169**:198–201.

[71] Perez LJ, Perera CL, Frias MT, Nunez JI, Ganges L, de Arce HD: A multiple SYBR Green I-based real-time PCR system for the simultaneous detection of porcine circovirus type 2, porcine parvovirus, pseudorabies virus and Torque teno sus virus 1 and 2 in pigs. J Virol Methods 2012, **179**:233–241.

[72] Zheng LL, Wang YB, Li MF, Chen HY, Guo XP, Geng JW, Wang ZY, Wei ZY, Cui BA: Simultaneous detection of porcine parvovirus and porcine circovirus type 2 by duplex real-time PCR and amplicon melting curve analysis using SYBR Green. J Virol Methods 2013, **187**:15–19.

[73] Finlaison DS: Faecal viruses of dogs: an electron microscope study. Vet Microbiol 1995, **46**:295–305.

[74] Decaro N, Desario C, Beall MJ, Cavalli A, Campolo M, Dimarco AA, Amorisco F, Colaianni ML, Buonavoglia C: Detection of canine parvovirus type 2c by a commercially available in-house rapid test. Vet J 2010, **184**:373–375.

[75] Lopez de Turiso JA, Cortes E, Ranz A, Garcia J, Sanz A, Vela C, Casal JI: Fine mapping of canine parvovirus B cell epitopes. J Gen Virol 1991, **72(Pt 10)**:2445–2456.

[76] Larson LJ, Schultz RD: Do two current canine parvovirus type 2 and 2b vaccines provide protection against the new type 2c variant? Vet Ther 2008, **9**:94–101.

[77] Spibey N, Greenwood NM, Sutton D, Chalmers WS, Tarpey I: Canine parvovirus type 2 vaccine protects against virulent challenge with type 2c virus. Vet Microbiol 2008, **128**:48–55.

[78] Decaro N, Desario C, Billi M, Mari V, Elia G, Cavalli A, Martella V, Buonavoglia C: Western European epidemiological survey for parvovirus and coronavirus infections in dogs. Vet J 2011, 187:195-199.

[79] Decaro N, Buonavoglia C: Canine parvovirus-a review of epidemiological and diagnostic aspects, with emphasis on type 2c. Vet Microbiol 2012, 155:1-12.

[80] Csagola A, Varga S, Lorincz M, Tuboly T: Analysis of the full-length VP2 protein of canine parvoviruses circulating in Hungary. Arch Virol 2014, 159:2441-2444.

[81] Filipov C, Decaro N, Desario C, Amorisco F, Sciarretta R, Buonavoglia C: Canine parvovirus epidemiology in Bulgaria. J Vet Diagn Invest 2011, 23:152-154.

[82] Muz D, Oguzoglu TC, Timurkan MO, Akin H: Characterization of the partial VP2 gene region of canine parvoviruses in domestic cats from Turkey. Virus Genes 2012, 44:301-308.

[83] Maya L, Calleros L, Francia L, Hernandez M, Iraola G, Panzera Y, Sosa K, Pérez R: Phylodynamics analysis of canine parvovirus in Uruguay: evidence of two successive invasions by different variants. Arch Virol 2013, 158:1133-1141.

[84] Castro TX, Costa EM, Leite JP, Labarthe NV, Cubel Garcia RC: Partial VP2 sequencing of canine parvovirus (CPV) strains circulating in the state of Rio de Janeiro, Brazil: detection of the new variant CPV-2c. Braz J Microbiol 2010, 41:1093-1098.

[85] Chiang SY, Wu HY, Chiou MT, Chang MC, Lin CN: Identification of a novel canine parvovirus type 2c in Taiwan. Virol J 2016, 13:160.

[86] Meers J, Kyaw-Tanner M, Bensink Z, Zwijnenberg R: Genetic analysis of canine parvovirus from dogs in Australia. Aust Vet J 2007, 85:392-396.

Chapter 2

Canine Visceral Leishmaniasis in Brazil

Marcia Almeida de Melo, Raizza Barros Sousa Silva,
Laysa Freire Franco e Silva,
Beatriz Maria de Almeida Braz,
Jaqueline Maria dos Santos,
Saul José Semião Santos and
Paulo Paes de Andrade

Additional information is available at the end of the chapter

http://dx.doi.org/10.5772/65956

Abstract

Visceral leishmaniasis (VL) is a disease caused by a protozoon belonging to the genus *Leishmania*, and it is transmitted through the bite of sand flies. Endemic regions have widened, and canine visceral leishmaniasis (CVL) occurs mainly in the Mediterranean region and South America. There is no consensus on the risk factors associated with CVL, as results differ between the studied regions and countries. This chapter describes the main aspects of epidemiology, immunology, clinical signs, diagnosis treatment, and control of canine visceral leishmaniasis with emphasis on Brazil.

Keywords: neglected diseases, risk factors, zoonosis

1. Introduction

Leishmaniases is a disease caused by protozoan intracellular parasites called *Leishmania*, members of Trypanosomatidae family, and are endemics in 98 countries and territories.

More than 20 parasite species are involved in the three different clinical manifestation diseases in human beings: cutaneous leishmaniasis, mucosal leishmaniasis, and visceral leishmaniasis (kala-azar). Post-kala-azar dermal leishmaniasis (PKDL) is a complication of visceral leishmaniasis in a patient who has recovered from the disease. The vectors are sand flies, insects of medical and veterinary relevance, and different species involved in its transmission.

According to Pan American Health Organization (PAHO) and World Health Organization (WHO), there are more than 12 million people infected with leishmaniasis, and 350 million are at risk in the world. Cutaneous leishmaniases are concentrated in ten countries, four of which are in the Americas: Brazil, Colombia, Peru, and Nicaragua. Ninety percent of visceral leishmaniasis cases occur in Brazil, Ethiopia, India, Bangladesh, Sudan, and South Sudan. In the Americas, an average 60,000 cases of cutaneous and mucosal leishmaniasis and 4000 cases of visceral leishmaniasis are diagnosed annually, with a fatality rate of 7% (**Figure 1**) [1].

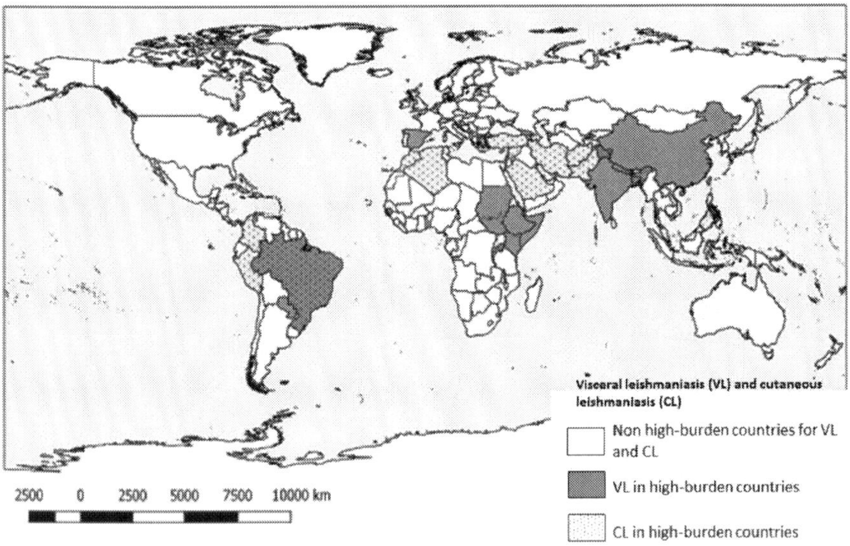

Figure 1. High-burden countries for both visceral and cutaneous leishmaniasis [2].

The aim of this chapter is to describe the main aspects of canine visceral leishmaniasis (CVL) with emphasis in Brazil.

2. Epidemiology

Visceral leishmaniasis (VL), also known as kala-azar, is a disease caused by an obligate intracellular protozoon belonging to the family Trypanosomatidae, genus *Leishmania*. It is transmitted through the bite of infected female sand flies and is widely distributed throughout the world (**Figure 2**). Currently, the World Health Organization considers VL as a neglected disease. An estimated 500,000 new cases of human VL occur per year [2]. Its epidemiology is extremely complex and depends on social and environmental variables, the species of *Leishmania* and the vector involved, and the behavior of reservoirs and hosts (**Table 1**).

http://dx.doi.org/10.5772/65956

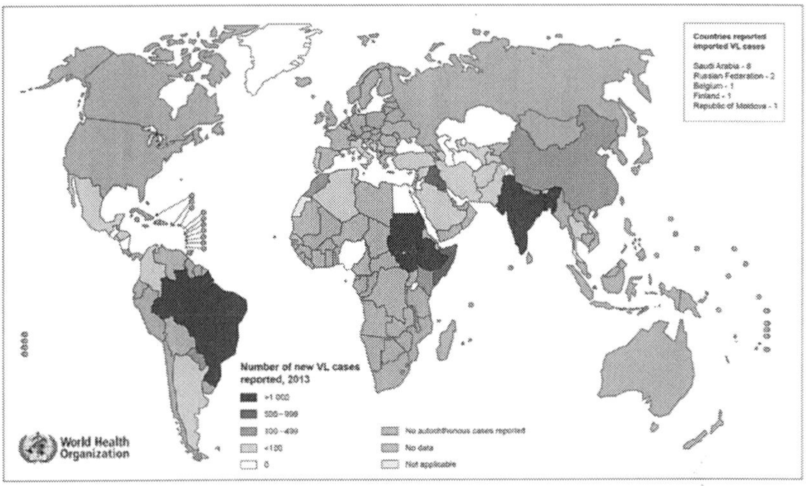

Figure 2. World distribution of human visceral leishmaniasis, 2013 [3].

Species	Region	Vector	Host/reservoir
Leishmania (L.) infantum	Old World:	*Phlebotomus perniciosus*	Humans
	Europe	*P. ariasi*	Dogs
	Asia		Wild canids
	Africa		
Leishmania (L.) donovani	Old World:	*P. argentipes*	Humans
	Asia	*P. orientalis*	
	Africa		
Leishmania (L.) infantum (syn. of *L. chagasi*)	New World:	*Lutzomyia longipalpis*	Humans
	South, Central, and North America	*Lu. cruzi*	Dogs
		Lu. forattinii	Wild canids
		Lu. evansi	Felines
			Marsupials

Adapted with permission from Refs. [4–6].

Table 1. Visceral leishmaniasis: species, region of occurrence, vectors, reservoirs, and mammal hosts.

2.1. Biological cycle

The life cycle of *L. infantum* is heteroxenic, that is, the parasite develops in two types of hosts: in the intermediate host, or the vector, the promastigote form develops in the insect gut; once in the definitive host, the amastigote develops as an obligate intracellular parasite in macrophages. During blood meals, vectors ingest macrophages containing *Leishmania* amastigotes, which will then multiply by binary division and differentiate into promastigotes. Promastigotes bind to the epithelium of the sand fly's esophagus or pharynx by the flagellum and then differentiate into metacyclic promastigotes, which is the infectious stage, unable to divide or bind to the midgut, remaining free in the digestive tract lumen [5, 7, 8].

Transmission occurs when the infected vector does a new blood meal and inoculates the infective form of *Leishmania* in the host. The sand fly regurgitates around 1000 metacyclics into the wound caused on the skin of a mammal [9]. When inoculated into the host organism, the parasite is phagocytized by the cells of mononuclear phagocytic system and loses the flagella. Next, it multiplies intensively by binary division, to the point of rupturing the host cell and releasing amastigotes, which then will be phagocytized by other cells and spread via blood and lymph to other tissues and organs [5, 7, 8].

2.2. Vector

VL vectors belong to the order Diptera, family Psychodidae, subfamily Phlebotominae, genus *Lutzomyia* in the new world, and *Phlebotomus* in the Old World. They can be found in tropical and subtropical regions of the planet and are popularly known as sand flies [10, 11].

Sand flies are small-sized, light brown-colored insects, with a coat of hair over their body. They measure between 2 and 3 mm [11]. Their flight range reaches about 150–300 m (diameter of 300–600 m), but may be longer depending on the species. They usually fly in small jumps and have crepuscular and nocturnal activity [12].

Hematophagy is a unique habit of females, who require blood for ovary maturation. Hence, they are able to transmit the disease. When females feed on blood, they hold their wings upright. A variety of animals has been identified as dietary hosts of sand flies, which have very eclectic feeding habits. Females lay eggs in moist soil, rich in organic matter [13].

2.3. Transmission

VL caused by *L. infantum* is a zoonosis and can display either wild or domestic transmission cycles depending on the eco-epidemiological conditions.

Leishmaniasis starts from the wild cycle, involving reservoirs such as wild canids and marsupials. As for the domestic cycle, the main reservoir is the domestic dog (Canis lupus familiaris) [7]. The link between wild and domestic cycles occurs when humans install dwellings on forest banks [14] and, most likely, because some wild reservoirs have synanthropic habits (**Figure 3**) [15].

Canine visceral leishmaniasis (CVL) is crucial from an epidemiological point of view, as it is more prevalent than human VL. Dogs exhibit high levels of subcutaneous parasites and high sensitivity to vectorial infection [12]. Additionally, between 50 and 60% of infected animals are asymptomatic [16]. It is estimated that three out of five asymptomatic positive dogs transmit the parasite to sand flies, and this transmission rate does not significantly change among symptomatic and asymptomatic groups of animals [17].

Other animals have been pointed as possible reservoirs, such as rodents [14] and cats, or accidental hosts, such as horses [18]. Adaptation of vectors to different animal species would be a favorable factor for VL transmission [19].

Figure 3. Eco-epidemiology of visceral leishmaniasis in Pará, Brazil. *L. infantum* parasite is maintained by a sand fly population (*Lutzomyia longipalpis*) and wild reservoirs (1). House invasion by sand flies near forests allows canine and human infection (2 and 3), where domestic dogs become the main source of the parasite. The same occurs in the urban areas. Solid lines indicate defined routes of transmission, and dashed lines represent possible transmission routes with other wildlife, and possibly with man, serving as a source of infection for sand flies [14].

There are other transmission forms of less epidemiological importance, such as blood transfusion and venereal transmission, apart from vertical transmission, which can be transplacental or transmammary [20–23]. Direct dog-to-dog transmission through bites or wounds has been suggested as responsible for sporadic CVL transmission as well [24, 25].

2.4. Geographic distribution and prevalence

In recent years, endemic regions have widened, and there has been a sharp increase in number of recorded cases of the disease. Still, it is believed that the impact of leishmaniasis on public health has been underestimated, since its notification is only mandatory in 32 of the 88 affected countries [2]. CVL is endemic in over 70 countries and occurs mainly in the Mediterranean region and South America [26]. Moreno and Alvar concluded that at least 2.5 million dogs are infected in only four countries in Southeastern portion of Western Europe, representing 16.7% of the canine population [27].

Franco et al. [28] developed a database of publications on the prevalence of LVC in Europe between the years 1971 and 2006. They found an overall prevalence of 23.2% and average of 10%, the highest taking place in Italy (17.7%), followed by France (8%), Portugal (7.3%), and Spain (5.9%) [28]. In another study, in 18 Portuguese cities with 3974 dogs, the overall prevalence of CVL was of 6.31% and ranged from 0.88% to 16.16% among cities, with high prevalence in inland regions [29].

In Croatia, seroprevalence of CVL ranges from 0 to 42.85% depending on the studied region [30]. In Cyprus, the seroprevalence of the disease in dogs had a ninefold increase compared to 10 years before, reaching 14.9% in 2010 [31]. The average seropositivity of dogs in Greece is 22.1%, and positive animals were found in 43 of 54 cities [32].

Seroprevalence of CVL in Southern Europe ranges from less than 5% to over 50% depending on the geographic region [33]. However, the prevalence of infection is significantly higher than both seroprevalence and apparent disease, due to sensitivity of serological techniques and because clinical signs usually only appear in less than half of the population, as in all endemic areas of the world [34].

Recent expansion to areas not previously endemic has been recorded in some parts of Europe, as northern Italy [35], in Germany [36], and in northern Spain [37]. The expansion of the CVL is associated with adjacent territories and often with global warming, which favors vector transmission, or with import of infected dogs to non-endemic areas, such as the United Kingdom and Poland [26, 33].

In Africa, VL transmission areas are located near forests [12]. Most of the information published in Africa have often been reports of human cases during epidemic situations, but information on the reservoirs are scarce [38]. In Algeria, the number of cases of CVL is on the rise, with a frequency of 35%, with 25% of positive dogs being asymptomatic [39]. Coastal regions of Northern Morocco are known to be endemic for canine and human VL. Moreover, several cases of CVL caused by *L. tropica* have been reported in areas where it is normally caused by *L. infantum* [40].

CVL exists in some Asian countries. It is endemic in Iran, with an average prevalence of 14.2%, and variations according to the region: 18.2% in northeast, 12.3% in the central region, and 4.4% in Southeastern Iran [41]. Another study, restricted to southwestern Iran, found a prevalence of 15.4% [42]. In China, VL is an anthropozoonosis which completes its life cycle in dogs, raccoons, coatis, and children [12]. The presence of *L. infantum* DNA in dogs, in Southwest China, ranged from 23.5 to 28.2% [43]. In the Western region, 51.9% of the dogs were positive in PCR, but only 36.8% of the dogs reacted in serology [44].

CVL has been expanding over the Americas. It currently occurs from southern Canada [45] and the United States [46, 47] to northern Argentina [48, 49]. In North America, it was first reported on hunting kennels of Foxhound dogs in New York, in 1999 [46]. Since then, CVL has been spreading and has so far been diagnosed in 18 North American and two Canadian states, totaling 58 kennels with positive Foxhounds, but with no reports of human cases. Isozyme characterization showed that the isolated agents from 46 Foxhounds are *Leishmania infantum* (MON1) [45]. High mortality and transmissibility rates associate with these foci in North America. It is believed that the disease was imported from Southern Europe due to dog travel history between these regions [50]. Direct transmission forms are associated with these outbreaks: from dog to dog, via placenta or during intercourse [22]. *L. infantum*, in these cases, remains capable of infecting sand flies and subsequently moving into a vertebrate host, which highlights the risk of vector-based transmission to humans [47].

There are few studies on CVL in Mexico, but human VL cases are constant. Rosete-Ortíz et al. analyzed skin lesions of 25 dogs by immunohistochemistry and PCR in a region where *Lutzomyia longipalpis* has been described and found *Leishmania* sp. in 60% of samples [51].

In Latin America, VL occurs in 12 countries, with 90% of cases concentrated in Brazil [11], distributed in all states of the Northeastern, Midwestern, and Southeastern regions, plus in Roraima, Tocantins, and Pará states, in the Northern region [14]. Among Brazilian cities and states, variation in prevalence of CVL is huge, ranging from 0.7% in Salvador, in the state of Bahia [52], to 51.6% in São Luis Island, Maranhão state [53], both in Northeast, where the disease is most prevalent.

CVL urbanization correlates with increasing global mobility [54] associated with demographic and ecological factors. In Latin America, especially in Brazil, Colombia, and Venezuela, migration and urbanization have contributed to the increase in American VL. In Brazil, one example is the rural exodus from the Northeast fields, causing thousands of people to migrate to cities like Fortaleza, Jacobina, João Pessoa, Natal, Petrolina, St. Louis, Sobral, Teresina, and Salvador. They then proceed to live in suburban areas with unsanitary conditions and malnutrition. Migrants often bring along their dogs and raise chicken and pigs around their homes, all ultimately serving as a feed source for the vector. According to Moreno et al., in urban areas, some factors favor the spread of *L. infantum*, such as the presence of waste and degraded areas around houses, lack of vector recognition by locals, and increased exposure to sand fly, especially in the early evening [55]. Thus, "ruralization" of the outskirts of the Northeastern cities has been identified as the cause of epidemics detected in the region [56], although this same process may not be necessary for entry of visceral leishmaniasis in other cities, as happened in the Southeast of the country.

The state of Rio Grande do Sul, Brazil, was considered CVL-free until 2008, when a case of an autochthonous canine was reported in São Borja [57]. In the same year, *Lutzomyia longipalpis* was found in urban areas of San Borja, a species not previously reported in the Southern region [58]. Since then, new cases were reported: 24 in dogs and 10 in humans between 2008 and 2014. Of these 34 reports, 20 were autochthonous, showing that it is an emerging zoonosis in the state [59]. It is believed that the disease has entered the state via Santo Tomé city in Argentina, which borders San Borja and already had reports of the disease [57, 59].

Prevalence of CVL in the world varies widely, and such variation also applies to different locations within the same city, suggesting that different ecosystems favor maintenance of vectors in different manners [60]. As noted by Azevedo et al. [61] and Belo et al. [62], results of studies on prevalence are influenced by various factors such as the region and population studied, the diagnostic method, as well as the sample used.

Migration of humans and their pets, disorderly occupation, poor living conditions, deforestation, and climate change associated with vector-adaptive capacity are some of the causes of the global urbanization of leishmaniasis [54].

The best example of the phenomenon of urbanization of zoonotic visceral leishmaniasis is happening in Brazil [54, 63]. VL has invaded urban centers and large capitals with no previous record of autochthonous cases [12]. Epidemiological data show the suburbanization and urbanization of visceral leishmaniasis, highlighting the outbreaks in Rio de Janeiro (RJ), Belo Horizonte (MG), Aracatuba (SP), Santarém (PA), Corumbá (MS), Teresina (PI), Natal (RN), São Luís (MA), Fortaleza (CE), Camaçari (BA), and more recently, occurrence of epidemics in the municipalities of Três Lagoas (MS), Campo Grande (MS), and Palmas (TO) [11].

The prevalence of human VL caused by *L. infantum* has decreased where living standards improved [2], showing that one of the most important interventions may be socioeconomic development and improved nutrition of children [8], but CVL cases have increased.

2.5. Risk factors

There is no consensus on the risk factors associated with CVL, as results differ between the studied Brazilian regions and between countries.

In Croatia, risk factors were sex (male), age (the two most prevalent groups comprise dogs between 3 and 4 years old and between 6 and 7 years old), and location (dogs in some cities are more likely to acquire the disease) [30]. In Spain, seroprevalence was also found to have bimodal age distribution, but the age groups were between 1 and 2 years and between 7 and 8 years; infection is also related to outdoor rearing [64]. In Portugal, risk factors are outdoor rearing, age (over two years), short fur, pure breeds, and location (dogs in the hinterlands are more likely to be affected) [29].

In Brazil, Belo et al. [62] conducted a systematic review of the literature on risk factors associated with CVL, and the variables that showed significant association with infection were short hair, pure breed, rearing restricted to house surroundings, and the presence of green areas adjacent to the house. The occurrence of CVL was also associated with the presence of poultry in domestic environment, free-living dogs, sex (male), and age greater than 1 or 2 years, although these associations were not statistically significant [62].

Another study in Brazil defined risk factors as outdoor rearing, contact with poultry, dogs living in rural areas, the presence of organic matter, the absence of environmental management, and proximity to forests [65]. As for dogs in the countryside, in an endemic area of Northeastern Brazil, the only identified risk factor found was sex, as male dogs were twice as likely to develop the disease [66].

3. Immunology

The components of the immune system act in a complex and coordinated manner to prevent entry and survival of foreign agents in the body. The first line of defense is the innate immunity, which responds immediately and unspecifically to a range of pathogens, and further presents them to the constituents of adaptive immunity when needed. Adaptive immunity will then generate a specific response and develop memory cells against such antigen. Performance of these defense mechanisms can control infection and ensure the least possible damage to host tissues.

In visceral leishmaniasis, the result of the relationship between parasite and host is determined by complex factors involving saliva components of the vector insect, agent-secreted surface proteins, and different responses produced by the host [67]. Leishmaniasis can be considered an immune-mediated disease, considering the parasite's ability to alter the immune system [68]. It ultimately promotes inhibition of immune response by either stimulating the development of regulatory T cells [69] or exerting some degree of control over the complement system, exploring its opsonic properties to facilitate adherence with phagocytic cells and preventing their lytic effects through the action of gp63 glycoprotein expressed on the parasite surface [70]. Thus, infection outcome depends on the parasite's capacity of developing evasion mechanisms to escape from host responses and remain unharmed in the cytoplasm of phagocytic cells.

It has been documented that resistance to infection by *L. infantum* in dogs is characterized by the absence of clinical signs, low levels of anti-*Leishmania* antibodies, in vitro lymphocyte proliferative response, and delayed hypersensitivity response to skin antigens. Progression of the infection, however, relates to exaggerated humoral response and cellular immune depression, consequently bringing up an onset of clinical signs [71–74]. Asymptomatic animals also exhibit lower parasitism, whereas the symptomatic generally carry high parasite load in different tissues such as the skin, bone marrow, spleen, liver, and lymph node [73–76].

3.1. Innate response

Through antigen presentation, the cells of innate immunity stimulate the acquired response. Antigen-presenting cells have receptors that recognize pathogen-associated molecular patterns (PAMPs) expressed by the parasite. Among these, the Toll-like receptor (TLR) is one of the most studied. Stimulation of these receptors culminates with the activation of signaling pathways in infected cells, which results in induction of antimicrobial genes and inflammatory cytokines (IL-12, TNF) while increasing the ability of cells to present antigen. Thus, pathogen recognition by TLR receptors helps conducting adaptive immune response against the presented antigen [77].

Expression of TLR genes in dogs infected with *L. infantum* varies according to the different tissues and stages of infection. As the disease progresses, there is a significant decrease in transcription for TLR3, TLR4, TLR9, IL-17, IL-22, and FoxP3 in lymph nodes. In spleen samples, decreased transcription for TLR4 and IL-22 has been observed when infected groups were compared with controls. In the skin, upregulation was observed only for TLR9 and FoxP3 in early stages of infection, as well as downregulation for TLR3 and TLR9 in later stages. Decrease in transcription of TLRs, Th17, and FoxP3 cytokines is suggestive of silent establishment of infection [78]. In peripheral blood samples from infected dogs, the highest expression of TLR2 and its receptor CD11b (CR3) by monocytes correlates with reduced parasitic load and higher resistance to leishmaniasis [79].

Studies have shown that both inlet and survival of *Leishmania* spp. within macrophages can occur from prior infection of neutrophils, which are recruited as a normal response to insect bites. The agent reaches the interior of macrophages when they phagocyte apoptotic bodies of previously infected neutrophils, where it can survive and multiply. This parasite escape mechanism is called "Trojan Horse" [80].

The main effector mechanism involved in protective immune response against *Leishmania* spp. is the activation of macrophages by IFN-γ and tumor necrosis factor alpha (TNF-α) stimulation, with consequent stimulation of nitric oxide synthesis, which is required for effective destruction of the pathogen and for controlling the spread of infection in dogs [81–83].

3.2. Acquired response

3.2.1. Lymphocytes

The major subpopulations of lymphocytes are CD4+ T cells (Th1, producing IFN-γ and TNF-α; Th2, secreting IL-4, IL-5, and IL-13; and Th17, producing IL-17 and IL-22) and CD8+ T cells. Antigen-presenting cells submit *Leishmania* spp. antigens to CD4+ T cells via MHC class II, and because the agent is an intracellular parasite, there may be presentation via MHC class I with activation of CD8+ T cells as well [69].

The role of CD4+ T cells in the response to visceral leishmaniasis (VL) has not been fully elucidated. Research carried out so far points to a mixed response (Th1/Th2) during infection [84, 85]. It is reported, however, that control of infection depends on Th1 cells that activate macrophages, promoting elimination of intracellular parasites [86], whereas Th2 cells direct the immune system toward humoral response and negatively regulate cellular immunity, promoting Th1 cell anergy [87].

CD8+ cells constitute a significant population in cellular immunity against canine visceral leishmaniasis (CVL), outnumbering CD4+ cells in the dermis [87]. They play an important role in resistance to infection. Guerra et al. [88] associated phenotypic changes with tissue parasitism in the spleen and skin of infected dogs. They noticed that the high frequency of CD8+-circulating lymphocytes is directly related to low splenic parasitism and that there is a negative correlation between CD8+ T cells with skin parasite density, indicating that this cell type relates to resistance against LVC [88].

The regulatory role of FOXP3+ CD4+ T cells in canine VL has not been fully elucidated; however, reduction of Treg cell percentage in peripheral blood of infected dogs has been observed [89]. Silva et al. [90] reported increased production of IL-10 by splenic Treg cells of dogs with LV, along with decrease in the total number of T cells when compared to healthy dogs. The findings suggest that Treg cells are a major source of IL-10 in the spleen and participate in the modulation of immune response, while a small percentage of these cells in infected dogs may be related to persistent immune activation [90].

T-cell exhaustion (CD4+ and CD8+ cells) in peripheral blood of dogs with LV, followed by reduction of the expression of cytokines (such as IFN-γ), was recently demonstrated. This phenomenon, called cell exhaustion, is mainly mediated by high expression of programmed death protein ("programmed cell death 1," PD-1) and may be related to the strong immunosuppression observed in advanced stages of the disease, corresponding to increase in symptomatology of VL [86].

The role of B cells in CVL is unclear. However, increases in CD21+ B cell, CD4+ T cell, and CD8+ T cell levels are frequently reported, as for the clinical asymptomatic form. Lower frequency of B cells and monocytes in the bloodstream is an important marker of severe disease, while increased levels of CD8 + T cells appear to be the most important phenotypic feature for asymptomatic clinical presentation [74]. Among symptomatic animals, decrease in CD21+ B-cell count associates with decreased CD4+ T cells, which does not occur with asymptomatic or disease-free control animals, suggesting that the decay of immunity in leishmaniasis may be related to decrease in the CD4+ T-cell population [91].

Although the relationship between a pattern of anti-*Leishmania* humoral response and resistance or susceptibility to LV is not well defined [92], immunoglobulin profile may appear as a biomarker for monitoring clinical prognosis and tissue parasitic density, as it is associated with the progression of clinical signs and increase of parasites in lymphoid organs [73]. Reis et al. reported increase in IgG1 levels associated with the asymptomatic clinical presentation and poor tissue parasitic load, while the symptomatic clinical form would be characterized by high parasitic load and high levels of anti-*Leishmania* IgG, IgG2, IgM, IgA, and IgE [73, 74]. Teixeira-Neto et al. reported similar results, with symptomatic dogs demonstrating high levels of anti-*Leishmania* IgG2 antibodies, whereas the asymptomatic showed higher IgG1 titles, when compared to IgG2 [93]. Production of IFN-γ is associated with the production of IgG1 [94], which may indicate a Th1 response profile.

3.2.2. *Cytokines and chemokines*

Cytokine patterns for CVL have not been well established. Studies are inconclusive, so the pattern of immune response associated with resistance or susceptibility in infected animals is yet to be established.

One of the first studies on cytokine profiling in CVL was performed by Pinelli et al. [71], who observed high levels of IL-2 and tumor necrosis factor alpha (TNF-α) in asymptomatic dogs, compared with symptomatic ones, which suggests a role of these cytokines in resistance to *L. infantum* in dogs naturally or experimentally infected. Since then, much research has been done in order to elucidate the cytokine profiles found in various tissue compartments of infected dogs. They revealed contrasting cytokine profiles among different tissues, indicating that the immune response in LVC occurs in an organ-specific manner [71].

Profile of cytokines in peripheral blood mononuclear cells (PBMC) culture from asymptomatic dogs experimentally infected with *L. infantum* shows a predominantly Th1 response, mediated by expression of IL-2, IFN-γ, and IL-18 and very low IL-4 expression [95]. De Lima [96] dosed IL-6 and TNF-α in serum of symptomatic dogs naturally infected with *L. infantum* and uninfected control dogs. TNF-α was found in similar levels in dogs of both groups, yet IL-6 showed statistically higher levels in dogs with active VL than healthy dogs, thus demonstrating to be a good indicator for the disease [96].

After evaluating expression of cytokines in spleen cells from dogs naturally infected with *L. infantum*, Lage et al. suggested that CVL is characterized by a mixed response: with production of cytokines types I and II; involvement of IFN-y and IL-10 and a positive correlation between IL-10 levels and progression of parasitic load or clinical manifestations of the disease; and correlation between IFN-γ and increased parasitic intensity of the spleen [84]. In whole blood, the increase of IL-10 has been associated with detection of parasite DNA [97]. Do Nascimento et al., in accordance, state that in CVL there is increase in expression of pro-inflammatory cytokines IFN-γ and TNF-α in dogs with low splenic parasitism, while dogs with higher parasitic load show an increase in IL-10 [98].

Souza [99] states that asymptomatic dogs have low dermal parasitism and exhibit a mixed pattern of immune response, with simultaneous increase of type I (IFN-γ and TNF-α) and type II (IL-5 and IL-13) cytokines, but predominance of type I response. According to the author, increased and simultaneous expression of IFN-γ and TNF-α in the skin of infected dogs enables the speculation that these mediators are closely involved with protection mechanisms during CVL, since these cytokines increased in the skin of animals with the asymptomatic clinical form. Increased expression of IL-5 and IL-13 in the skin of healthy dogs and negative correlation of the latter with clinical disease progression were also observed. Furthermore, high simultaneous expression of IFN-γ and IL-13 was found in asymptomatic dogs, indicating the role of IL-13 in establishing milder clinical forms [99].

Regarding cytokine profile in the bone marrow, Quinnell et al. [94] reported that expression of mRNA for IL-10, IL-4, and IL-18 was not elevated in infected dogs. However, some infected dogs had detectable expression of mRNA for IL-4 significantly correlated with more severe clinical signs. Moreover, expression of mRNA for IL-13 was not detected either in control or in infected dogs, and unlike in human infection, immunosuppressive activity of IL-10 was not observed in CVL [94].

Dogs infected with *L. infantum* exhibit significant decrease in expression of mRNAs for IL-10, IL-17, TNF-α, IFN-γ, and iNOS in liver tissue. Deficiency in IL-17 mRNA expression was evident in the symptomatic dogs compared to the asymptomatic. Reduction in cytokine expression results in decreased iNOS expression and therefore higher parasite load. The increase in IL-17 expression in the liver of asymptomatic dogs and its correlation with elevated expression of iNOS indicates a protective role of that cytokine in canine infection by *L. infantum* [67]. However, Michelin et al. [100] reported increased TNF-α, IL-4, and IL-10 levels in the liver of infected dogs compared to healthy dogs. In addition, the association between TNF-α levels and an increase in parasitic load in the spleen suggests its importance in the evolution of the infection process [100].

Another subject lacking clarification is the participation of chemokines and their receptors in resistance or susceptibility to LVC. Knowledge surrounding the role of these modulators in response to *L. infantum* infection is critical, considering that the interaction between cytokines and chemokines may regulate the immune response against the parasite, activating and recruiting immune cells to areas of infection.

Menezes-Souza et al. [101] analyzed the expression of CCL2, CCl4, CCL5, CCL13, CCL17, CCL21, CCL24, and CXCL8 chemokines in the skin of 35 dogs naturally infected by *L. infantum*, comparing cutaneous parasitism and clinical manifestations of the disease. Increase of the parasitic load correlated with increased expression of CCL2, CCL4, CCL5, CCL21, and CXCL8. On the other hand, CCL24 expression negatively correlated with parasitism [101].

After connecting clinical findings in naturally infected dogs with liver and spleen parasitism and expression levels for cytokines, chemokines, and their receptors, Albuquerque [67] showed that symptomatic dogs exhibit low expression of these modulators, alongside lower inflammatory response, and higher parasite load—primarily in the liver—than asymptomatic animals. CXCL10 was the only chemokine found at a much higher concentration in both the liver and the spleen of symptomatic animals. It also positively correlated with clinical score. The author indicates that expression profiles of hepatic and splenic chemokines and their receptors are essential for induction of correct cell inflammatory profile, as it has potential to contain the infection and the disease. Impaired cell migration facilitates replication of the parasite and development of CVL symptoms [67].

Understanding of the immune response in canine visceral leishmaniasis may reveal the factors involved with the onset and severity of clinical signs and the damage to host tissues. Additionally, it takes place as an indispensable tool for development of an effective vaccine.

4. Clinical signs and symptoms

Canine visceral leishmaniasis presents a clinical picture ranging from asymptomatic to classical symptomatic cases (**Figure 4**). Infected animals without clinical signs comprise an

Figure 4. Classical symptomatic case with emaciation, thickened skin, cutaneous lesions, exfoliative dermatitis, fur loss, and cutaneous ulcers.

alarming 40–80% of all cases, for asymptomatic dogs are a major source of infection for sand flies, and owners naturally resist to elimination of their animals [102].

At first, kala-azar signals can be rather discrete and easily confused with other diseases. Animals may have discrete lesions on the edge of the ears and slight changes in blood profile (mild anemia and/or thrombocytopenia).

Clinical leishmaniasis may appear quickly after infection or within two years. Classic canine kala-azar is characterized by thickened skin, cutaneous lesions, intermittent fever, appearance shift and fur loss, periorbital alopecia, hepatosplenomegaly, akinesia, diarrhea, onychogryphosis (nail growth; **Figure 5**), and nosebleeding. Partial paralysis of hindquarters is often seen in the final stage of the disease.

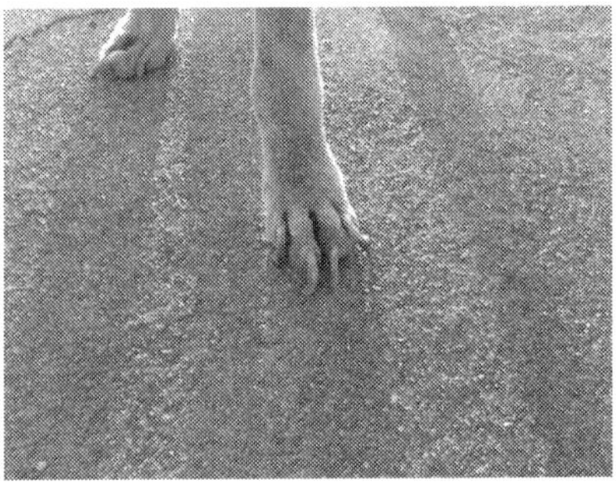

Figure 5. Onychogryphosis.

The most common skin lesions in dogs with kala-azar are exfoliative dermatitis (generalized, regional, or localized); ulcerative dermatitis, onychogryphosis, and papular dermatitis.

Cutaneous ulcers are usually located on the ear margins and have been attributed to local trauma and/or vasculitis, pressure points (**Figure 6**), limbs, and mucocutaneous junctions. Focal or multifocal nodular forms have a high amastigote load and may indicate either inefficient or strong cellular immunity by the host [101, 103–105].

Ocular disease occurs in CVL, with anterior uveitis being the most common ocular manifestation, characterized by conjunctivitis, blepharitis, periocular alopecia, exophthalmia, keratitis, keratoconjunctivitis sicca, anterior uveitis, glaucoma, and retinal detachment [106].

The nosebleeding (epistaxis) occurs due to thrombocytopenia and is often confused with ehrlichiosis, a bacterial disease transmitted by ticks, which in many cases might associate with leishmaniasis. In endemic areas, it is advisable that any diagnosis of ehrlichia or anaplasma in dogs must be accompanied by differential diagnosis of kala-azar.

In general, dogs in endemic areas are poly-infected and malnourished, particularly stray dogs or those who frequently wander on the streets, leading to a plurality of overlapping clinical pictures. Among other conditions, furfuraceous flaking due to scabies, weight loss as consequence of other infections or lymphomas, and autoimmune diseases, such as systemic lupus erythematosus, often confuse diagnosis since clinical signs are usually not pathognomonic. Therefore, differential diagnosis must be a concern for the small animal clinician [107, 108].

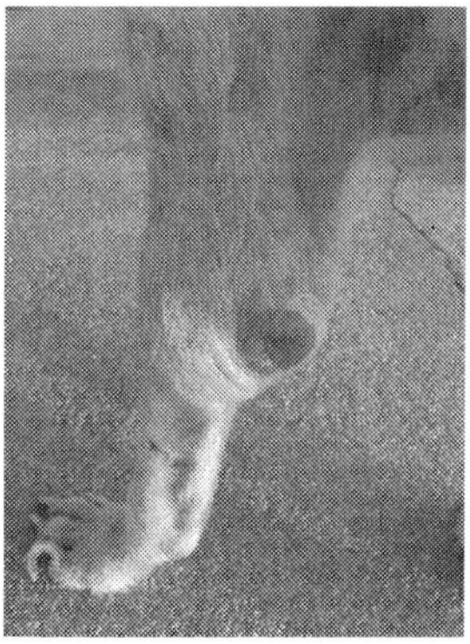

Figure 6. Cutaneous ulcer located on a pressure point.

5. Treatment

Treating seropositive dogs for canine visceral leishmaniasis (CVL) is a controversial practice in Brazil and, above all, not recommended by the World Health Organization, mainly because it does not lessen the importance of the dog as a reservoir, and utilizes drugs used in human treatment of visceral leishmaniasis (VL) [109]. Nevertheless, European countries legally established treatment since the twentieth century [110].

Frequent usage of these drugs in veterinary clinics may select resistant parasites due to variation in sensitivity of leishmania species, in addition to providing low parasiticide effect, thus interfering negatively in human treatment [111]. The lack of success for parasitological cure occurs mainly because it is an intracellular parasite and is located in less vascularized tissues where it can be difficult to obtain therapeutic doses, such as the vitreous body [112].

Many studies have been conducted in order to find an effective treatment for CVL, but drugs currently available are still inefficient, only allowing temporary remission of clinical signs. Besides, some have a high cost and produce toxic effects. Pentavalent antimonies (glucamine antimoniate—Glucantime® or sodium stibogluconate—Pentostam®) are widely used in CVL therapy protocols because they usually produce faster clinical remission but are often combined with allopurinol, since they do not prevent relapses [34]. A variety of drugs, such as amphotericin B, pentamidine isethionate, ketoconazole, fluconazole, miconazole, itraconazole, has been used either isolated or in combination and produced different results [111].

5.1. Chemotherapy

Due to the possibility of parasitic resistance to first-choice drugs, chemotherapeutic treatment options are limited. In addition, due to the high cost and long-term use, chemotherapy becomes an undesirable option for owners.

In veterinary medicine, the first choice as chemotherapy for treatment of CVL is allopurinol, a leishmaniostatic drug that acts by inhibiting leishmania growth through DNA modification. It has low cost, but parasite resistance to it remains unknown. Allopurinol is however the only drug recommended by the World Health Organization, especially since it is little used for treatment of human leishmaniasis [113].

Antimonials are leishmanicidal drugs that hinder promastigote metabolism by inhibiting glycolytic activity. These are drugs of choice for human treatment of VL but are chemically similar to those that have been used in therapeutic protocols in dogs. Their toxicity and efficacy are related to the antimony content [113].

When assessing the therapeutic efficacy of Glucantime alone or associated with an antigenic extract of *L. braziliensis* in experimentally infected asymptomatic dogs, reduction in parasite burden and antibody levels decreased, but actual cure was not achieved. Yet, upon using meglumine antimoniate and allopurinol—alone and in combination—in dogs naturally infected by *L. infantum*, symptomatic animals showed remission of clinical signs about 60 days after the treatment, accompanied by restoring of normal urinary and hematologic profiles. Despite clinical improvement of treated animals, however, the combination was not able to eliminate the parasite in all dogs. Thus, this protocol is not recommended, since treated dogs would go on as a source of infection for humans and other dogs, similar to results obtained using antimoniate, which promoted hematologic normalization and bone marrow recovery, but not parasitological cure, and to findings from another study wherein dogs showed increased parasitic load in lymph nodes [114–117].

In accordance with Ikeda-Garcia et al. [116], Manna et al. [118] monitored leishmania DNA load in infected dogs through real-time PCR, using a similar treatment protocol. Therapy resulted in clinical improvement accompanied by reduction in parasite burden, but even after a long period of treatment with allopurinol alone, parasites remained in tissues [116, 118].

Miltefosine is another drug used in human treatment that has been evaluated for canine treatment in recent years. It is a phospholipid antibiotic of broad spectrum with leishmanicide effect that improves the activation of both macrophages and T cells [119]. After evaluating efficacy of three treatment protocols for dogs naturally infected with *L. infantum*, significant clinical improvement was observed and later progressed to full recovery and humoral stimulation, but did not assure cure, on the contrary, progressively increased parasite load. Hence, miltefosine alone is not recommended for treatment. The study also investigated the profile of major cytokines involved in CVL healing process, and findings showed an increase of IFN-γ within 90 days after treatment, when the parasite load in bone marrow aspirates also decreased. There was a decrease of IL-4, indicating possible resolution of the infection and efficacy of the treatment, since this cytokine is a good marker for the occurrence of active disease [112, 120].

Some studies seek to associate chemotherapeutic treatment to immunomodulatory drugs, which plays a role in therapeutic protocols by controlling clinical signs and in prevention protocols by enhancing the immune cell-mediated response through activation of macrophages via helper T cells, in order to destroy phagocytized microorganisms. Domperidone is a receptor antagonist of dopamine D2 that has been used as well. When orally given to naturally infected dogs, results showed a reduction of clinical signs and titers of anti-leishmania antibodies [121]. In Spain, Sabaté et al. [122] used a treatment protocol with domperidone in seronegative dogs in an area with high prevalence of *L. infantum* and later exposed the animals to the vector. After 21 days of treatment, the treated group had lower antibody titer, confirming that implementation of treatment with this drug reduces the risk of developing CVL [122]. Nevertheless, further clinical trials are needed to determine the optimal dose, as well as posttreatment investigations to check for parasitological cure, since the drug is a good choice for infected animals with kidney failure due to its route of administration, and it is a low-cost option [121].

Amphotericin B is a broad-spectrum macrolide antibiotic produced by actinomycete *Streptomyces nodosus* [123], available for usage in VL treatment. It has a powerful leishmanicide effect and works by binding to esters on the parasite's plasma membrane [111]. According to Solano-Gallego et al., this drug has disadvantages concerning its route of administration (IV) and its nephrotoxic potential; however, its use results in less failures and recurrences [34]. Additionally, it is frequently used in the treatment of both human and canine VL for being inexpensive.

Athanasiou et al. [124] investigated the effectiveness of aminosidine sulfate, a leishmanicide antibiotic of the aminoglycoside class, in the treatment of dogs naturally infected. They observed reduction in clinical signs and parasite density, in antibody titers through indirect immunofluorescence, and in prevalence of positive dogs using PCR 3 months after the end of the experiment. These findings can be explained by direct action of the drug as well as activation of cellular immunity. However, more studies are required to prove its therapeutic efficacy for CVL, especially considering its affordable price to owners and market availability [124].

Despite the research on efficacy of different classes of medications for CVL treatment, no great progress has been done regarding toxicity or parasitological cure, highlighting the necessity for evaluation of new formulations and medicaments to be used exclusively for treatment of CVL.

5.2. Immunochemotherapy

Immunotherapy can be an effective addition to chemotherapy, as it induces effector immune response faster than the isolated use of chemotherapeutic drugs [125]. In a trial, the combination of N-methyl meglumine antimoniate (Glucantime®) and lyophilized recombinant vaccine Leish-110f®, together with adjuvant monophosphoryl lipid A plus (MPL-SE®) to treat symptomatic animals naturally infected with *L. infantum*, was effective in reducing parasitic load. This combination enables reduction of chemotherapy doses, consequently lowering the risks of toxicity and death. However, the use of multiple recombinant antigens could provide better results [126].

Borja-Cabréra et al. [127] evaluated the efficacy of immunotherapeutic vaccine Leishmune® administered alone, both in commercial formulation enriched with saponin and in laboratory formulation, compared to its use in combination with amphotericin B and allopurinol, in dogs naturally infected with *L. infantum*, but found no significant differences between the vaccine formulations tested. As for the type of treatment, the group submitted to immunochemotherapy with saponin-enriched Leishmune® vaccine abolished, not only the symptoms but also the latent infection condition, curing the dogs [127].

Joshi et al. [128] used Balb/c mice in order to verify in vivo therapeutic potential of the first generation of vaccines with dead *L. donovani* antigen (KDL) combined with adjuvant (MPL-A) and leishmanicidal chemotherapy such as cisplatin and sodium stibogluconate (SSG) and then compared those to isolated chemotherapy and immunotherapy. In animals treated with vaccine associated to SSG, there was a 98.50% reduction in serum parasitic levels, but the study also noted that the use of any of the chemotherapeutic associated with vaccination resulted in direct parasite elimination by drug activity and activation of the immune cell-mediated response. They concluded that immunochemotherapy protocols may be effective, but further studies are needed in different animal models in order to better understand the immune response [128].

5.3. Immunotherapy

Immunotherapy is often considered for CVL prevention and control as a preferable alternative to euthanasia, due to the absence of a low-toxicity chemotherapy treatment and increasing resistance. According to Grandoni [129], vaccines are regarded as the best tool for eradication of the disease, particularly because it reduces the incidence of new cases, considering that the immune system fails to efficiently control the infection as it mediates a weak protective cell response. This relates to a dichotomy between the trigger of a Th1 response related to resistance and a Th2 response associated with susceptibility to infection, increased parasitic load, and strong but ineffective humoral immune response [129].

According to Joshi et al., an effective vaccine must induce a strong and long-lasting Th1 response as to prevent the initial establishment of infection: by definition, a prophylactic vaccine [128]. However, when it comes to a disease caused by an obligate intracellular protozoan, the aim is to at least control progression to severe disease and prevent transmission from host to vector, hindering maintenance of the epidemiological cycle.

So far, vaccines formulated for CVL include dead parasites, protein components or parasite subunits, purified cell fractions, vector salivary recombinant proteins, and viral particles that encode parasite's virulence factors and plasmid DNA [130]. Leishmune® (Zoetis Animal Health) was the first vaccine approved for commercial use in Brazil, in 2003. However, in 2014 the Ministry of Agriculture, Livestock, and Supply (MALS) halted manufacturing and marketing licenses due to problems in phase III. It consists of a glycoprotein antigen that binds recombinant fucose mannose which was able to stimulate good cellular immune response, decreasing IL-4, and activate CD4+ T cells, producing TNF-α and IFN-γ, important cytokines in resistance [131, 132]. Accordingly, a study done by Borja-Cabrera et al. found that the vaccine induced a long-lasting protective effect of humoral and cellular immunity, along with disappearance of clinical signs and parasitemia [133].

Leish-Tec® (Hertape), a vaccine comprising recombinant protein A2 as antigen in adjuvant saponin QuilA, continues to be marketed in Brazil since 2008, when it was recorded by the MALS. It has been demonstrated to offer partial protection against *L. infantum* in Beagle dogs, producing good humoral and cellular immune responses, high levels of anti-protein A2 IgG and IgG2, and IFN-γ [134, 135].

Despite increasing progress in production of vaccines against CVL in Brazil and worldwide, there is much to be improved regarding the induction of durable and efficient cellular and humoral immune response. Moreover, new affordable vaccines ought to be produced for the population, since those available in the market so far are expensive and not viable for use in public health.

6. Diagnosis

The clinical diagnosis of CVL is challenging, as signs are usually not specific for the disease; laboratory assays are therefore of paramount importance. Moreover, as the dog is considered to be the major reservoir in Brazil, serological assays constitute the basis to identify infected dogs and to direct public health actions aiming the disease control.

The indirect immunofluorescence antibody test (IFAT) was established as the standard serodiagnosis in public health programs more than 40 years ago. It was later substituted by an ELISA based on crude leishmania antigens [111] and more recently by a recombinant ELISA and an immunochromatographic rapid test for detecting K26/K39-reactive antibodies in canine sera [136, 137]. According to the recommendation from the Brazilian Ministry of Health, sera collected from dogs in seroepidemiologic surveys as part of the Leishmaniasis Control Program are first screened using the fast immunochromatographic assay (known as dual-path platform (DPP)), and the positive samples are retested in the recombinant ELISA. The approach undoubtedly speeds up the screening [138], but due to the DPP low sensitivity in cases of sera from asymptomatic dogs [137, 139, 140], a sizable set of infected dogs possibly remains in the endemic areas, jeopardizing the effectiveness of control actions.

Private laboratories used to rely in a single result obtained by the use of a commercial recombinant ELISA kit (ELISA S7) registered at the Brazilian Ministry of Agriculture, Livestock, and Supply for the diagnosis of CVL (www.biogene.ind.br). More recently, some fast immunochromatographic assays started to be used, such as the Alere assay [141], but high costs preclude their adoption in the routine diagnosis. A recombinant K39-based ELISA developed for research use only (http://www.inbios.com/kalazar-detect-elisa-system-for-visceral-leishmaniasis-intl/) has also seen some use in routine commercial CVL diagnosis.

The official recombinant ELISA assay and the ELISA S7 have similar sensitivity and specificity indexes and perform equally well in the identification of seropositive, supposedly infected dogs. They also do not display significant rates of positive reactions in cases of vaccinated dogs. Other serological assays, e.g., the direct agglutination (DAT), are not easily available in Brazil and were never adopted in private or public labs. Although new antigens have been described in the last years (e.g., [142]), their commercial use in serodiagnosis is still uncertain. Although available as commercial kits (http://www.genesig.com/products/9332?gclid=CNXfgtXc_c4CFcoHkQodMm4Ctw), PCR assays are seldom used and have limited application for the routine diagnosis.

As the existing recombinant ELISA assays have high sensitivity and specificity even in the serodiagnosis of CVL in asymptomatic dogs, one could argue against the need of new laboratory tests. However, claims of cross reactions with babesiosis and other common canine infectious diseases [138] continue to stir dissatisfaction.

In conclusion, no diagnostic breakthroughs have been described and no innovative technology was introduced in the market in the last years, and there are no evidences that this scenario will be changed in the near future.

7. Epidemiological surveillance, preventive, and control measures

Visceral leishmaniasis control activities focus on reducing morbidity and mortality through early diagnosis and treatment of human cases, monitoring and euthanasia of seropositive dogs and sand fly population control via entomological surveillance such as chemical control. Additionally, they include education and health activities that involve joint actions aiming to improve population's quality of life, such as provision of basic sanitation and proper trash disposal [143].

In Brazil, the Visceral Leishmaniasis Control Program (VLCP) recommends surveillance, preventive and control measures of human and canine visceral leishmaniasis [11]. Epidemiological surveillance aims to reduce mortality and morbidity rates through early diagnosis and treatment of human cases and to reduce the risk of transmission by controlling reservoirs and vector populations. Surveillance comprises entomological surveillance of human and canine cases. What is set at national level for epidemiological surveillance of VL, emphasizing canine population, is described in the following paragraphs.

Through epidemiological analysis of VL in the state or municipality, transmission areas are classified in areas with VL cases or silent areas (without cases). Areas with cases are those with record of a first confirmed case, those with sporadic, moderate, and intense transmission and those undergoing outbreaks. Silent areas or areas without cases are classified as vulnerable (receptive and unreceptive) or not vulnerable.

Entomological surveillance aims to gather quantitative and qualitative information about *L. longipalpis* and/or *L. cruzi* vectors. This is a task for the state and/or municipal health departments, which must work integratedly in order to optimize resources and effectiveness of sand fly control activities.

In regard to dogs, surveillance focuses on suspect canine cases (symptomatic animals in endemic or outbreak areas) and on those confirmed by (a) laboratory criteria, symptomatic

http://dx.doi.org/10.5772/65956

and positive in serological and/or parasitological test; (b) clinical and epidemiological criteria, symptomatic from endemic or outbreak areas without diagnostic confirmation; and (c) infected dogs, asymptomatic and positive in serological and/or parasitological test. When a canine case is identified, delimitation of the area to be investigated is among the surveillance actions to be taken. In those areas, active search for symptomatic dogs must be carried out for parasitological examination, and if the agent is found, serological survey of all animals in the area must be done in order to evaluate local prevalence and to implement appropriate measurements.

Sample and census serosurveys must be performed as monitoring activity. The sample serological survey must be carried out in silent and receptive municipalities with *L. Longipalpis* and in those with moderate and intense transmission. The census serological survey must be performed in urban areas of silent and receptive municipalities with canine population smaller than 500 animals, in urban areas of moderate or intense transmission and rural areas in any transmission situation. The aim of a census survey is to control disease by identification of infected dogs in order to perform euthanasia and assess prevalence. It must be performed annually for at least three consecutive years, even without notification of new confirmed human cases.

Preventive measures regarding canine population are (a) control of errant canine population, (b) donation of dogs after performing negative serological tests, (c) vaccination against CVL, (d) the use of fine mesh screen at individual or collective kennels, and (e) the use of collars impregnated with deltamethrin 4%.

Canine reservoir control measures consist in euthanasia of seropositive dogs and/or positive in parasitological tests, besides disposal of the bodies in accordance with the provisions of RDC Resolution No. 33, of February 25, 2003, from the National Agency for Health Surveillance.

In Brazil, since 2000, as part of the decentralization process undergone by the National Health Foundation (FUNASA), the states' federal district and municipalities became responsible for operating assistance, epidemiology, and disease control activities. They now receive almost all movable property, allocated in all federal units and more than 26,000 servers, and resources for maintenance of the transferred responsibilities [144]. However, despite the decentralization and recommendations by VLCP, the action taken toward canine reservoirs have been mainly restricted to areas of human case occurrences, that is, after a human case confirmation, canine serological survey follows in the surrounding area, and later, euthanasia of animals found seropositive in screening (DPP®) and confirmatory (EIE-Biomanguinhos) tests. Individual preventive measures such as vaccination, coverage of kennel doors and windows with mesh screen, and collars impregnated with deltamethrin 4% are restricted to animals whose owners have relatively high economic standard.

Several factors hinder the fulfillment of activities imposed by VLCP. Some of them are the lack of federal funding; insufficient staff to perform activities related to VL and other endemic diseases; prioritization for control of other endemic diseases such as dengue, Zika, and chikungunya; expansion of transmission areas; and interference of veterinarians, animal owners; and nongovernmental organizations (NGOs) regarding euthanasia of reservoirs, as such procedure generates controversy regarding its control efficacy, although recommended in Brazil [145].

Acknowledgements

Funding provided by The National Science Center (Poland) grant No. 2014/13/B/NZ4/03832.

Author details

Marcia Almeida de Melo[1,2*], Raizza Barros Sousa Silva[2], Laysa Freire Franco e Silva[2], Beatriz Maria de Almeida Braz[2], Jaqueline Maria dos Santos[3], Saul José Semião Santos[4] and Paulo Paes de Andrade[5]

*Address all correspondence to: marcia.melo@pq.cnpq.br

1 Federal University of Campina Grande (UFCG), Patos, Brazil

2 Postgraduate Program Veterinary Medicine/UFCG, Patos, Brazil

3 Postgraduate Program in Animal and Tropical Science/Federal Rural University of Pernambuco (UFRPE), Recife, Brazil

4 Tiradentes University, Aracaju, Brazil

5 Federal University of Pernambuco (UFPE), Recife, Brazil

References

[1] PAHO/WHO. Key facts on Neglected Infectious Diseases. Leishmaniasis [Internet]. 2014. Available from: file://D:/Pessoal/Downloads/2014-cha-leishmaniasis-fact-sheet.pdf [Accessed: 2016-07-17]

[2] WHO. Leishmaniasis [Internet]. 2016. Available from: http://www.who.int/mediacentre/factsheets/fs375/en/ [Accessed: 2016-07-17]

[3] WHO. Status of endemicity of visceral leishmaniasis, worldwide, 2013 [Internet]. 2015. Available from: file:///D:/Pessoal/Downloads/Leishmaniasis_Burden_distribution_VL_CL_2013.pdf [Accessed: 2016-07-17]

[4] Pita-Pereira D, Cardoso MAB, Alves CR, Brazil RP, Britto C. Detection of natural infection in *Lutzomyia cruzi* and *Lutzomyia forattinii* (Diptera: Psychodidae: Phlebotominae) by *Leishmania infantum chagasi* in an endemic area of visceral leishmaniasis in

Brazil using a PCR multiplex assay. Acta Tropica. 2008; 107:66–69. DOI: 10.1016/j.actatropica.2008.04.015

[5] Carreira JCA, Magalhães MAFM, Silva AVM. The geospatial approach on eco-epidemiological studies of leishmaniasis. In: Claborn DM. (Ed.) Leishmaniasis–trends in epidemiology, diagnosis and treatment. InTech; 2014. p. 125–145. DOI: 10.5772/57067

[6] Akhoundi M, Kuhls K, Cannet A, Votýpka J, Marty P, Delaunay P, Sereno D. A historical overview of the classification, evolution, and dispersion of leishmania parasites and sandflies. PLoS Negl Trop Dis. 2016; 10(3):1–40. DOI: 10.1371/journal.pntd.0004349

[7] Harhay MO, Olliaro PL, Costa DL, Costa CHN. Urban parasitology: visceral leishmaniasis in Brazil. Trends Parasitol. 2011; 27(9):403–409. DOI:10.1016/j.pt.2011.04.001

[8] Ready PD. Epidemiology of visceral leishmaniasis. Clinical Epidemiology. 2014; 6:147–154. DOI: 10.2147/CLEP.S44267

[9] Schlein Y, Warburg A. Phytophagy and the feeding cycle of *Phlebotomus papatasi* (Diptera: Psychodidae) under experimental conditions. J Med Entomol. 1986; 23:11–15

[10] Ready PD. Biology of phlebotomine sand flies as vectors of disease agents. Annu Rev Entomol. 2013; 58:227–250. DOI: 10.1146/annurev-ento-120811-153557

[11] Brasil. Ministério da Saúde. Manual de vigilância e controle da leishmaniose visceral. 1th ed. Brasília (DF): Ministério da Saúde; 2014. 120 p.

[12] Troncarelli MZ, Carneiro DMVF, Langoni H. Visceral leishmaniosis: an old disease with continuous impact on public health. In: Lorenzo-Morales J. (Ed.) Zoonosis. InTech; 2012. p. 263–282. DOI: 10.5772/38680

[13] Dias FOP, Lorosa ES, Rebêlo JMM. Blood feeding source and peridomiciliation of *Lutzomyia longipalpis* (Lutz … Neiva, 1912) (Psychodidae, Phlebotominae). Cad Saúde Pública. 2003; 19(5):1373–1380. DOI: 10.1590/S0102-311X2003000500015.

[14] Lainson R, Rangel E. *Lutzomyia longipalpis* and the eco-epidemiology of American visceral leishmaniasis, with particular reference to Brazil—a review. Mem Inst Oswaldo Cruz. 2005;100:811–827. DOI: /S0074-02762005000800001

[15] Gontijo CMF, Melo MN. Visceral leishmaniasis in Brazil: current situation, challenges and perspectives. Rev Bras Epidemiol. 2004; 7(3):338–349. DOI: 10.1590/S1415-790X2004000300011

[16] Alvar J, Cañavate C, Molina R, Moreno J, Nieto J. Canine leishmaniasis. Adv Parasitol. 2004; 57:1–88. DOI: 10.1016/S0065-308X(04)57001-X

[17] Molina R, Amela C, Nieto J, San-Andrés M, González F, Castillo JA, Lucientes J, Alvar J. Infectivity of dogs naturally infected with *Leishmania infantum* to colonized *Phlebotomus perniciosus*. Trans R Soc Trop Med Hyg. 1994;88:491–493.

[18] Gramiccia M. Recent advances in leishmaniosis in pet animals: Epidemiology, diagnostics and anti-vectorial prophylaxis. Vet Parasitol. 2011; 181:23–30. DOI: 10.1016/j.vetpar.2011.04.019

[19] Dantas-Torres F, de Brito ME, Brandão-Filho SP. Seroepidemiological survey on canine leishmaniasis among dogs from an urban area of Brazil. Vet Parasitol. 2006; 140(1–2): 54–60. DOI: 10.1016/j.vetpar.2006.03.008

[20] de Freitas E, Melo MN, da Costa-Val AP, Michalick MSM. Transmission of *Leishmania infantum* via blood transfusion in dogs: potential for infection and importance of clinical factors. Vet Parasitol. 2006; 137:159–167. DOI: 10.1016/j.vetpar.2005.12.011

[21] Silva FL, Oliveira RG, Silva TM, Xavier MN, Nascimento EF, Santos RL. Venereal transmission of canine visceral leishmaniasis. Vet Parasitol. 2009; 160(1–2):55–59. DOI: 10.1016/j.vetpar.2008.10.079.

[22] Boggiatto PM, Gibson-Corley KN, Metz K, Gallup JM, Hostetter JM, Mullin K, Petersen CA. Transplacental transmission of *Leishmania infantum* as a means for continued disease incidence in North America. PLoS Negl Trop Dis. 2011; 5(4):e1019. DOI: 10.1371/journal.pntd.0001019.

[23] Oliveira VVG, Alves LC, Silva Junior VA. Transmission routes of visceral leishmaniasis in mammals. Cienc Rural [online]. 2015; 45(9):1622–1628. DOI: 10.1590/0103-8478cr20141368.

[24] Schantz PM, Steurer FJ, Duprey ZH, Kurpel KP, Barr SC, Jackson JE, Breitschwerdt EB, Levy MG, Fox JC. Autochthonous visceral leishmaniasis in dogs in North America. J Am Vet Med Assoc. 2005; 226:1316–1322.

[25] Naucke TJ, Amelung S, Lorentz S. First report of transmission of canine leishmaniosis through bite wounds from a naturally infected dog in Germany. Parasit Vectors. 2016; 9(1):256. DOI: 10.1186/s13071-016-1551-0.

[26] Kaszak I, Planellas M, Dworecka-Kaszak B. Canine leishmaniosis—an emerging disease. Ann Parasitol. 2015; 61(2):69–76.

[27] Moreno J, Alvar J. Canine leishmaniasis: epidemiological risk and the experimental model. Trends Parasitol. 2002; 18(9):399–405

[28] Franco AO, Davies CR, Mylne A, Dedet JP, Gállego M, Ballart C, Gramiccia M, Gradoni L, Molina R, Gálvez R, Morillas-Márquez F, Barón-López S, Pires CA, Afonso MO, Ready PD, Cox J. Predicting the distribution of canine leishmaniasis in western Europe based on environmental variables. Parasitology. 2011; 138(14):1878–1891. DOI: 10.1017/S003118201100148X

[29] Cortes S, Vaz Y, Neves R, Maia C, Cardoso L, Campino L. Risk factors for canine leishmaniasis in an endemic Mediterranean region. Vet Parasitol. 2012; 189(2–4):189–196. DOI: 10.1016/j.vetpar.2012.04.028

[30] Zivicnjak T, Martinković F, Marinculić A, Mrljak V, Kucer N, Matijatko V, Mihaljević Z, Barić-Rafaj R. A seroepidemiologic survey of canine visceral leishmaniosis among apparently healthy dogs in Croatia. Vet Parasitol. 2005; 131,35–43. DOI: 10.1016/j.vetpar.2005.04.036

[31] Mazeris A, Soteriadou K, Dedet JP, Haralambous C, Tsatsaris A, Moschandreas J, Messaritakis I, Christodoulou V, Papadopoulos B, Ivovic V, Pratlong F, Loucaides F, Antoniou M. Leishmaniases and the Cyprus paradox. Am J Trop Med Hyg. 2010; 82(3): 441–448. DOI: 10.4269/ajtmh.2010.09-0282

[32] Ntais P, Sifaki-Pistola D, Christodoulou V, Messaritakis I, Pratlong F, Poupalos G, Antoniou M. Leishmaniases in Greece. Am J Trop Med Hyg. 2013; 89(5):906–915. DOI: 10.4269/ajtmh.13-0070

[33] Maia C, Cardoso L. Spread of *Leishmania infantum* in Europe with dog travelling. Vet Parasitol. 2015; 213(1–2):2–11. DOI: 10.1016/j.vetpar.2015.05.003

[34] Solano-Gallego L, Koutinas A, Miró G, Cardoso L, Pennisi MG, Ferrer L, Bourdeau P, Oliva G, Baneth G. Directions for the diagnosis, clinical staging, treatment and prevention of canine leishmaniosis. Vet Parasitol. 2009; 165(1–2):1–18. DOI: 10.1016/j.vetpar.2009.05.022

[35] Maroli M, Rossi L, Baldelli R, Capelli G, Ferroglio E, Genchi C, Gramiccia M, Mortarino M, Pietrobelli M, Gradoni L. The northward spread of leishmaniasis in Italy: evidence from retrospective and ongoing studies on the canine reservoir and phlebotomine vectors. Trop Med Internat Health. 2008; 13(2):256–264. DOI: 10.1111/j.1365-3156.2007.01998.x

[36] Naucke TJ, Menn B, Massberg D, Lorentz S. Sandflies and leishmaniasis in Germany. Parasitol Res. 2008; 103(1):S65–S68. DOI: 10.1007/s00436-008-1052-y

[37] Miró G, Checa R, Montoya A, Hernández L, Dado D, Gálvez R. Current situation of *Leishmania infantum* infection in shelter dogs in northern Spain. Parasit Vectors. 2012; 5:60. DOI: 10.1186/1756-3305-5-60

[38] Boakye DA, Wilson M, Kweku M. A review of leishmaniasis in west Africa. Ghana Med J. 2005; 39(3):94–97.

[39] Harrat Z, Belkaid M. Les leishmanioses dans l'Algérois. Données épidémiologiques. Bull Soc Pathol Exot. 2003; 96(3):212–214.

[40] Rhajaoui M, Nasereddin A, Fellah H, Azmi K, Amarir F, Al-Jawabreh A, Ereqat S, Planer J, Abdeen Z. New clinicoepidemiologic profile of cutaneous leishmaniasis, Morocco, Emerg Infect Dis. 2007; 13(9):1358–1360. DOI: 10.3201/eid1309.070946

[41] Mohebali M, Hajjaran H, Hamzavi Y, Mobedi I, Arshi S, Zarei Z, Akhoundi B, Naeini KM, Avizeh R, Fakhar M. Epidemiological aspects of canine visceral leishmaniosis in the Islamic Republic of Iran. Vet Parasitol. 2005; 129:243–251. DOI: 10.1016/j.vetpar.2005.01.010

[42] Mahshid M, Baharak A, Iraj S, Sina K, Javad K, Mehdi B. Seroprevalence of canine visceral leishmaniasis in southeast of Iran. J Parasit Dis. 2014; 38(2):218–222. DOI: 10.1007/s12639-012-0226-9

[43] Shang L, Peng W, Jin H, Xu D, Zhong N, Wang W, Wu Y, Liu Q. The prevalence of canine *Leishmania infantum* infection in Sichuan Province, southwestern China detected by real time PCR. Parasit Vectors 2011; 4:173. DOI: 10.1186/1756-3305-4-173

[44] Wang J, Ha Y, Gao C, Wang Y, Yang Y, Chen H. The prevalence of canine *Leishmania infantum* infection in western China detected by PCR and serological tests. Parasit Vectors. 2011; 4:69. DOI: 10.1186/1756-3305-4-69

[45] Duprey ZH, Steurer FJ, Rooney JA, Kirchhoff LV, Jackson JE, Rowton ED, Schantz PM. Canine Visceral Leishmaniasis, United States and Canada, 2000–2003. Emerg Infect Dis. 2006; 12(3): 440–446. DOI: 10.3201/eid1205.050811

[46] Gaskin AA, Schantz P, Jackson J, Birkenheuer A, Tomlinson L, Gramiccia M, Levy M, Steurer F, Kollmar E, Hegarty BC, Ahn A, Breitschwerdt EB. Visceral leishmaniasis in a New York foxhound kennel. J Vet Intern Med. 2002;16(1):34–44.

[47] Schaut RG, Robles-Murguia M, Juelsgaard R, Esch KJ, Bartholomay LC, Ramalho-Ortigao M, Petersen CA. Vectorborne Transmission of *Leishmania infantum* from Hounds, United States. Emerg Infect Dis. 2015; 21(12):2209–2212. DOI: 10.3201/eid2112.141167

[48] Salomón OD, Sinagra A, Nevot MC, Barberian G, Paulin P, Estevez JO, Riarte A, Estevez J. First visceral leishmaniasis focus in Argentina. Mem Inst Oswaldo Cruz. 2008; 103(1): 109–111. DOI: 10.1590/S0074-02762008000100018

[49] Salomón OD, Quintana MG, Bruno MR, Quiriconi RV, Cabral V. Visceral leishmaniasis in border areas: clustered distribution of phlebotomine sand flies in Clorinda, Argentina. Mem Inst Oswaldo Cruz. 2009; 104(5):801–804. DOI: 10.1590/S0074-02762009000500024

[50] Petersen CA, Barr SC. Canine leishmaniasis in North America: emerging or newly recognized? Vet Clin North Am Small Anim Pract. 2009; 39(6):1065–1074. DOI: 10.1016/j.cvsm.2009.06.008.

[51] Rosete-Ortíz D, Berzunza-Cruz MS, Salaiza-Suazo NL, González C, Treviño-Garza N, Ruiz-Remigio A, Gudiño-Zayas ME, Beltrán-Silva S, Romero-Zamora JS, Ugarte-Soto A, Rivas-Sánchez B, Becker I. Canine leishmaniasis in Mexico: the detection of a new focus of canine leishmaniasis in the state of Guerrero correlates with an increase of human cases. Bol Med Hosp Infant Mex. 2011; 68(2):88–93.

[52] Barboza DCPM, Leal DC, Souza BMPS, Carneiro AJB, Gomes Neto CMB; Alcânatara ACD, Julião FS, Moura SAB, Peralva LMP, Ferreira F, Franke CR. Canine visceral

leishmaniasis epidemiological survey in three sanitary districts of Salvador, Bahia State, Brazil, Rev Brasil Saúde Prod Anim. 2009; 10(2):434–447.

[53] Silva-Abreu AL, Lima TB, Macedo AA, Moraes-Júnior FJ, Dias EL, Batista ZS, Calabrese KS, Moraes JLP, Rebêlo JMM, Guerra RMSNC. Seroprevalence, clinical and biochemical data of dogs naturally infected by Leishmania and phlebotominae sandfly fauna in an endemic area in São Luis Island, Maranhão State, Brazil, Rev Bras Parasitol Vet. 2008; 17:197–203.

[54] WHO. Urbanization: an increasing risk factor for leishmaniasis. Weekly epidemiological record. 2002; 44(77):365–370. Available from: http://www.who.int/wer [Accessed: 2016-07-17]

[55] Moreno EC, Melo MN, Genaro O, Lambertucci JR, Serufo JC, Andrade AS, Antunes CM, Carneiro M. Risk factors for *Leishmania chagasi* infection in an urban area of Minas Gerais State. Rev Soc Bras Med Trop. 2005;38(6):456–463. DOI:/S0037-86822005000600002

[56] Cesse EAP, Carvalho EF, Andrade PP, Ramalho WM, Luna LKS. The organization of urban areas and expansion of kala-azar. Revista Brasileira de Saúde Materno Infantil. 2001; 1(2):167–176.

[57] Azevedo JSC, Esmeraldino AT, Ávila VPF, Witz MI, Fischer CDB, Tartarotti AL. Autochthonous canine visceral leishmaniasis in the municipality of São Borja, Rio Grande do Sul, Brazil: a case report. Veterinária em Foco. 2009; 7(1):52–61.

[58] Souza GD, Santos ES, Andrade Filho JD. The first report of the main vector of visceral leishmaniasis in America, *Lutzomyia longipalpis* (Lutz & Neiva) (Diptera: Psychodidae: Phlebotominae), in the state of Rio Grande do Sul, Brazil, Mem Inst Oswaldo Cruz. 2009; 104(8):1181–1182. DOI: 10.1590/S0074-02762009000800017

[59] Souza APL, Jesus JR, Teixeira MC. Retrospective study of visceral leishmaniasis epidemiology in Rio Grande do Sul: a literature review. Veterinária em Foco. 2014; 11(2): 112–118.

[60] França-Silva JC, Costa RT, Siqueira AM, Machado-Coelho GLL, Costa CA, Mayrink W, Vieira EP, Costa JS, Genaro O, Nascimento E. Epidemiology of canine visceral leishmaniasis in the endemic área of Montes Claros Municipality, Minas Gerais state, Brazil, Vet Parasitol. 2003; 111:161–173. DOI: 10.1016/S0304-4017(02)00351-5

[61] Azevedo MAA, Dias AKK, Paula HB, Perri SHV, Nunes CM. 2008. Evaluation of canine visceral leishmaniasis in Poxoréo, Mato Grosso, Brazil. Rev Bras Parasitol Vet. 17(3): 123–127. DOI: 10.1590/S1984-29612008000300001

[62] Belo VS, Struchiner CJ, Werneck GL, Barbosa DS, Oliveira RB, Neto RGT, Silva ES. A systematic review and meta-analysis of the factors associated with *Leishmania infantum*

infection in dogs in Brazil. PLoS Negl Trop Dis. 2013; 7(4):e2182. DOI: 10.1371/journal.pntd.0002182

[63] Costa CHN. Characterization and speculations on the urbanization of visceral leishmaniasis in Brazil. Cad Saúde Pública. 2008; 24(12):2959–2963. DOI: 10.1590/S0102-311X2008001200027

[64] Gálvez R, Miró G, Descalzo MA, Nieto J, Dado D, Martín O, Cubero E, Molina R. Emerging trends in the seroprevalence of canine leishmaniosis in the Madrid region (central Spain). Vet Parasitol. 2010; 169:327–334. DOI: 10.1016/j.vetpar.2009.11.025

[65] Costa AP, Costa FB, Soares HS, Ramirez DG, Araújo AC, Ferreira JGS, Tonhosolo R, Dias RA, Gennari SM, Marcili A. Environmental Factors and Ecosystems Associated with Canine Visceral Leishmaniasis in Northeastern Brazil. Vector Borne Zoonotic Dis. 2015; 15(12):765–774. DOI: 10.1089/vbz.2015.1866

[66] Silva RBS, Mendes RS, Santana VL, Souza HC, Ramos CPS, Souza AP, Andrade PP, Melo MA. Epidemiological aspects of canine visceral leishmaniasis in the semi-arid region of Paraiba and analysis of diagnostic techniques. Pesq Vet Bras. 2016;36(7):625–629.

[67] Albuquerque TDR. Correlation between immune response and the clinical manifestations of Canine Visceral Leishmaniasis. Natal: UFRN; 2013.

[68] Barbieri CL. Immunology of canine leishmaniasis. Parasite Imunol. 2006; 28:329–337. DOI: 10.1111/j.1365-3024.2006.00840.x

[69] Abbas AK, Lichtman AH, Pillai S. Imunologia celular e molecular. 7th ed. Rio de Janeiro: Elsevier Editora Ltda; 2011.592 p. DOI: 10.1084/gen.20090209

[70] Brittingham A, Morrison CJ, McMaster WR, McGuire BS, Chang KP, Mosser DM. Role of the Leishmania surface protease gp63 in complement fixation, cell adhesion, and resistance to complement-mediated lysis. J Immunol. 1995; 155(6):3102–3111.DOI: 10.1016/0169-4758(95)80054-9

[71] Pinelli E, Killick-Kendrick R, Wagenaar J, Bernadina W, Real G, Ruitenberg J. Cellular and humoral immune responses in dogs experimentally and naturally infected with *Leishmania infantum*. Infect Immun. 1994; 62(1):229–235.

[72] Santos-Gomes GM, Rosa R, Leandro C, Cortes S, Romão P, Silveira H. Cytokine expression during the outcome of canine experimental infection by *Leishmania infantum*. Vet Immunol Immunopathol. 2002; 88(1–2):21–30. DOI: 10.1016/S0165-2427(02)00134-4

[73] Reis AB, Teixeira-Carvalho A, Vale AM, Marques MJ, Giunchetti RC, Mayrink W, Guerra LL, Andrade RA, Corrêa-Oliveira R, Martins-Filho OA. Isotype patterns of immunoglobulins: hallmarks for clinical status and tissue parasite density in Brazilian dogs naturally infected by *Leishmania (Leishmania) chagasi*. Vet Immunol Immunopathol. 2006; 112(3–4):102–116. DOI: 10.1016/j.vetimm.2006.02.001

[74] Reis AB, Martins-Filho OA, Teixeira-Carvalho A, Giunchetti RC, Carneiro CM, Mayrink W, Tafuri WL, Corrêa-Oliveira R. Systemic and compartmentalized immune response in canine visceral leishmaniasis. Vet Immunol Immunopathol. 2009; 128(1-3):87–95. DOI: 10.1016/j.vetimm.2008.10.307

[75] Giunchetti RC, Martins-Filho OA, Carneiro CM, Mayrink W, Marques MJ, Tafuri WL, Corrêa-Oliveira R, Reis AB. Histopathology, parasite density and cell phenotypes of the popliteal lymph node in canine visceral leishmaniasis. Vet Immunol Immunopathol. 2008; 121(1-2):23–33. DOI: 10.1016/j.vetimm.2007.07.009

[76] Giunchetti RC, Mayrink W, Carneiro CM, Corrêa-Oliveira R, Martins-Filho OA, Marques MJ, Tafuri WL, Reis AB. Histopathological and immunohistochemical investigations of the hepatic compartment associated with parasitism and serum biochemical changes in canine visceral leishmaniasis. Res Vet Sci. 2008; 84(2):269–277. DOI: 10.1016/j.rvsc.2007.04.020

[77] Janeway CA, Medzhitov R. Innate immune recognition. Annu Rev Immunol. 2002; 20:197–216. DOI: 10.1146/annurev.immunol.20.083001.084359

[78] Hosein S, Rodríguez-Cortés A, Blake DP, Allenspach K, Alberola J, Solano-Gallego L. Transcription of toll-like receptors 2, 3, 4 and 9, FoxP3 and Th17 cytokines in a susceptible experimental model of canine *Leishmania infantum* infection. PLoS ONE. 2015;10(10). DOI: 10.1371/journal.pone.0140325

[79] Amorim IF, Silva SM, Figueiredo MM, Moura EP, Castro RS, Lima TK, Gontijo Nde F, Michalick MS, Gollob KJ, Tafuri WL. Toll receptors type-2 and CR3 expression of canine monocytes and its correlation with immunohistochemistry and xenodiagnosis in visceral leishmaniasis. PLoS ONE. 2011; 6(11). DOI: 10.1371/journal.pone.0027679

[80] Laskay T, Van Zandbergen G, Solbach W. Neutrophil granulocytes as host cells and transport vehicles for intracellular pathogens: apoptosis as infection-promoting factor. Immunobiology. 2008; 213(3–4):183–191. DOI: 10.1016/j.imbio.2007.11.010

[81] Liew FY, Wei XQ, Proudfoot L. Cytokines and nitric oxide as ejector molecules against parasitic infections. Philos Trans R Soc Lond B. 1997; 352:1311–1315. DOI: 10.1098/rstb.1997.0115

[82] Vouldoukis I, Drapier JC, Nüssler AK, Tselentis Y, Da Silva OA, Gentilini M, Mossalayi DM, Monjour L, Dugas B. Canine visceral leishmaniasis: successful chemotherapy induces macrophage antileishmanial activity via the l-arginine nitric oxide pathway. Antimicrob Agents Chemother. 1996; 40(1):253–256.

[83] Zafra R, Jaber JR, Pérez-Ecija RA, Barragán A, Martínez-Moreno A, Pérez J. High iNOS expression in macrophages in canine leishmaniasis is associated with low intracellular parasite burden. Vet Immunol Immunopathol. 2008; 123(3–4):353–3599. DOI: 10.1016/j.vetimm.2008.02.022

[84] Lage RS, Oliveira GC, Busek SU, Guerra LL, Giunchetti RC, Corrêa-Oliveira R, Reis AB. Analysis of the cytokine profile in spleen cells from dogs naturally infected by

Leishmania chagasi. Vet Immunol Immunopathol. 2007; 115(1–2):135–145. DOI: 10.1016/j.vetimm.2006.10.001

[85] Strauss-Ayali D, Baneth G, Jaffe CL. Splenic immune responses during canine visceral leishmaniasis. Vet Res. 2007; 38(4):547–564. DOI: 10.1051/vetres:2007015

[86] Esch KJ, Juelsgaard R, Martinez PA, Jones DE, Petersen CA. PD-1-mediated T cell exhaustion during visceral leishmaniasis impairs phagocyte function. J Immunol. 2013; 191(11):5542–5550. DOI: 10.4049/jimmunol.1301810

[87] Papadogiannakis EI, Koutinas AF. Cutaneous immune mechanisms in canine leishmaniosis due to *Leishmania infantum*. Vet Immunol Immunopathol. 2015; 163(3–4):94–102. DOI: 10.1016/j.vetimm.2014.11.011

[88] Guerra LL, Teixeira-Carvalho A, Giunchetti RC, Martins-Filho OA, Reis AB, Corrêa-Oliveira R. Evaluation of the influence of tissue parasite density on hematological and phenotypic cellular parameters of circulating leukocytes and splenocytes during ongoing canine visceral leishmaniasis. Parasitol Res, 2009; 104(3):611–622. DOI: 10.1007/s00436-008-1237-4

[89] Cortese L, Annunziatella M, Palatucci AT, Rubino V, Piantedosi D, Di Loria A, Ruggiero G, Ciaramella P, Terrazzano G. Regulatory T cells, cytotoxic T lymphocytes and a TH1 cytokine profile in dogs naturally infected by *Leishmania infantum*. Res Vet Sci. 2013; 95(3):942–949. DOI: 10.1016/j.rvsc.2013.08.005

[90] Silva KL, de Andrade MM, Melo LM, Perosso J, Vasconcelos RO, Munari DP, Lima VM. CD4+FOXP3+ cells produce IL-10 in the spleens of dogs with visceral leishmaniasis. Vet Parasitol. 2014; 202(3–4):313–318. DOI: 10.1016/j.vetpar.2014.03.010

[91] Bourdoiseau G, Marchal T, Magnol JP. Immunohistochemical detection of *Leishmania infantum* in formalin-fixed, paraffin-embedded sections of canine skin and lymph nodes. J Vet Diagn Invest. 1997; 9(4):439–440.

[92] Day MJ. Immunoglobulin G subclass distribution in canine leishmaniosis: a review and analysis of pitfalls in interpretation. Vet Parasitol. 2007; 147(1–2):2–8. DOI: 10.1016/j.vetpar.2007.03.037

[93] Teixeira-Neto RG, Giunchetti RC, Carneiro CM, Vitor RW, Coura-Vital W, Quaresma PF, Ker HG, de Melo LA, Gontijo CM, Reis AB. Relationship of Leishmania-specific IgG levels and IgG avidity with parasite density and clinical signs in canine leishmaniasis. Vet Parasitol. 2010; 169(3–4):248–257. DOI: 10.1016/j.vetpar.2010.01.023

[94] Quinnell RJ, Courtenay O, Shaw MA, Day MJ, Garcez LM, Dye C, Kaye PM. Tissue cytokine responses in canine visceral leishmaniasis. J Infect Dis. 2001; 183(9):1421–1424. DOI: 10.1086/319869

[95] Chamizo C, Moreno J, Alvar J. Semi-quantitative analysis of cytokine expression in asymptomatic canine leishmaniasis. Vet Immunol Immunopathol. 2005; 103(1–2):67–75. DOI: 10.1016/j.vetimm.2004.08.010

[96] de Lima VM, Peiro JR, Vasconcelos RO. IL-6 and TNF-a production during active canine visceral leishmaniasis. Vet Immunol Immunopathol. 2007; 115(1–2):189–193. DOI: 10.1016/j.vetimm.2006.10.003

[97] Boggiatto PM, Ramer-Tait AE, Metz K, Kramer EE, Gibson-Corley K, Mullin K, Hostetter JM, Gallup JM, Jones DE, Petersen CA. Immunologic Indicators of clinical progression during canine *Leishmania infantum* infection. Clin Vaccine Immunol. 2010; 17(2):267–273. DOI: 10.1128/CVI.00456-09

[98] do Nascimento PR, Martins DR, Monteiro GR, Queiroz PV, Freire-Neto FP, Queiroz JW, Morais Lima AL, Jeronimo SM. Association of pro-inflammatory cytokines and iron regulatory protein 2 (IRP2) with Leishmania burden in canine visceral leishmaniasis. PLoS ONE. 2013; 8(10):e73873. DOI: 10.1371/journal.pone.0073873

[99] Souza DM. The role of cytokines, chemokines and transcription factors in the immunopathology of skin in dogs naturally infected with *Leishmania (Leishmania)* chagasi, people with different clinical forms and cutaneous parasite densities. Belo Horizonte: Fundação Oswaldo Cruz, 2009.

[100] Michelin AF,Perri SH, De Lima VM. Evaluation of TNF-a, IL-4, and IL-10 and parasite density in spleen and liver of L. (L.) chagasi naturally infected dogs. Ann Trop Med Parasitol. 2011; 105(5):373–383. DOI: 10.1179/1364859411Y.0000000027.

[101] Menezes-Souza D, Guerra-Sá R, Carneiro CM, Vitoriano-Souza J, Giunchetti RC, Teixeira-Carvalho A, Silveira-Lemos D, Oliveira GC, Corrêa-Oliveira R, Reis AB. Higher expression of CCL2, CCL4, CCL5, CCL21, and CXCL8 chemokines in the skin associated with parasite density in canine visceral leishmaniasis. PLoS Negl Trop Dis. 2012; 6:e1566. DOI:10.1371/journal.pntd.0001566

[102] Madeira MF, Schubach AO, Schubach TM, Leal CA, Marzochi MCA. Identification of *Leishmania (Leishmania) chagasi* isolated from healthy skin of symptomatic and asymptomatic dogs seropositive for leishmaniasis in the municipality of Rio de Janeiro, Brazil. Braz J InfectDis 2004; 8:440–444. DOI: 10.1590/S1413-86702004000600008

[103] Cavalcanti A, Lobo R, Cupolillo E, Bustamante F, Porrozzi R. Canine cutaneous leishmaniasis is caused by neotropical *Leishmania infantum* despite of systemic disease: a case report. Parasitol Int. 2012; 61:738–740. DOI: 10.1016/j.parint.2012.05.002.

[104] Moura EP, Ribeiro RR, Sampaio WM, Lima WG, Alves CF, Melo FA, Melo MN,Tafuri WL,Tafuri WL,Michalick MSM. Histopathological and parasitological analysis of skin tissues biopsies from two distinct anatomical areas of the ears of dogs naturally infected with *Leishmania (Leishmania) chagasi.*Braz J Vet Pathol. 2008; 1:10–15.

[105] Levy E, Mylonakis ME, Saridomichelakis MN,Polizopoulou ZS, Psychogios V, Koutinas AF.Nasal and oral masses in a dog. Vet Clin Pathol. 2006; 35:115–118.

[106] Peña MT, Naranjo C, Klauss G, Fondevila D, Leiva M, Roura X, Davidson MG, Dubielzig RR. Histopathological features of ocular leishmaniosis in the dog. J Comp Pathol. 2008; 138(1):32–39.

[107] Cafarchia C, Gallo S, Danesi P, Capelli G, Paradies P, Traversa D, Gasser RB, Otranto D. Assessing the relationship between Malassezia and leishmaniasis in dogs with or without skin lesions. Acta Trop. 2008;107:25–29. DOI: 10.1016/j.actatropica.2008.04.008

[108] Tarantino C, Rossi G, Kramer LH, Perrucci S, Cringoli G, Macchioni G.*Leishmania infantum* and *Neospora caninum* simultaneous skin infection in a young dog in Italy. Vet Parasitol. 2001; 102:77–83.

[109] WHO. Control of the leishmaniasis. Geneve: WHO. (Technical Report, Series 793), 1990.

[110] Tesh RB. Control of zoonotic visceral leishmaniasis: is it time to change strategies? Am J Trop Med Hyg. 1995;52:287–292.

[111] Brasil. Ministério da Saúde. Manual de vigilância e controle da leishmaniose visceral, 1th ed. Brasília (DF): Ministério da Saúde; 2006. 120 p.

[112] Ciaramella P, Corona M. Canine Leishmaniasis: clinical and diagnostic aspects. Comp Cont Educ Pract Vet.2003; 25:358–368.

[113] WHO. Control of Leishmaniases. Geneva: WHO. (Technical Report Series 949), 2010.

[114] Melo MA, França-Silva JC, Azevedo EO, Tabosa IM, Costa RT, Genaro O, Mayrink W, Costa JO. Clinical Trial on the efficacy of the N-methyl glucamine associated to immunotherapy in dogs, experimentally infected with *Leishmania (Leishmania) chagasi.* Revue de MédecineVétérinaire. 2002;153(2):75–84.

[115] Ikeda-Garcia FA, Augusta F, Lopes RS, Marques FJ, Ciarlini PC, Lima VMF, Morinishi CK, Zanette MF, Perri SHV, Marcondes M. Clinical and parasitological evaluation of dogs naturally infected by *Leishmania (Leishmania) chagasi* submitted to treatment with meglumine antimoniate. Vet Parasitol. 2007; 143(3–4):254–259. DOI: 10.1016/j.vetpar.2006.08.019

[116] Ikeda-Garcia FA, Lopes RS, Marques FJ, de Lima VM, Morinishi CK, Bonello FL, Zanette MF, Perri SH, Feitosa MM. Clinical and parasitological evaluation of dogs naturally infected by *Leishmania (Leishmania) chagasi* submitted to treatment with meglumine antimoniate and allopurinol. Braz J Vet Res Anim Sci. 2010; 47(3):218–223. DOI: 10.526/cab.v11i2.2149

[117] Manna L, Corso R, Galiero G, Cerrone A Muzj P, Gravino AE. Long-term follow-up of dogs with leishmaniosis treated with meglumine antimoniate plus allopurinol versus miltefosine plus alopurinol. Parasit Vectors. 2015;8:289. DOI: 10.1186/s13071-015-0896-0

[118] Manna L, Reale S, Vitale F, Picillo E, Pavone LM, Gravino AE. Real-time PCR assay in Leishmania-infected dogs treated with meglumine antimoniate and allopurinol. Vet J. 2008; 177(2):279–282. DOI: 10.1016/j.tvjl.2007.04.013

[119] Soto J, Toledo J, Valda L, Balderrama M, Rea I, Parra R, Ardiles J, Soto P, Gomez A, Molleda F, Fuentelsaz C, Anders G, Sindermann H, Engel J, Berman J. Treatment of Bolivian mucosal leishmaniasis with miltefosine. Clin Infect Dis. 2007; 44(3):350–356. DOI: 10.1086/510588

[120] Andrade HM, Toledo VP, Pinheiro MB, Guimarães TM, Oliveira NC, Castro JA, Silva RN, Amorim AC, Brandão RM, Yoko M, Silva AS, Dumont K, Ribeiro ML Jr, Bartchewsky W, Monte SJ. Evaluation of miltefosine for the treatment of dogs naturally infected with *L. infantum* (=*L. chagasi*) in Brazil. Vet Parasitol. 2011; 181(2–4):83–90. DOI: 10.1016/j.vetpar.2011.05.009

[121] Gómez-Ochoa P, Castillo JA, Gascón M, Zarate JJ, Alvarez F, Couto CG. Use of domperidone in the treatment of canine visceral leishmaniasis: a clinical trial. Vet J. 2009; 179(2):259–263. DOI: 10.1016/j.tvjl.2007.09.014

[122] Sabaté D, Llinás J, Homedes J, Sust M, Ferrer L. A single-centre, open-label, controlled, randomized clinical trial to assess the preventive efficacy of a domperidone-based treatment programme against clinical canine leishmaniasis in a high prevalence area. Prevent Vet Med. 2014; 115(1–2):56–63. DOI: 10.1016/j.prevetmed.2014.03.010

[123] Noli C, Auxilia S.T. Treatment of canine Old World visceral leishmaniasis:a systematic review. Vet Dermatol. 2005; 16(4):213–232. DOI: 10.1111/j.1365-3164.2005.00460.x

[124] Athanasiou LV, Saridomichelakis MN, Kontos VI, Spanakos G, Rallis TS. Treatment of canine leishmaniosis with aminosidine at an optimized dosage regimen: a pilot open clinical trial. Vet Parasitol. 2013; 192(1–3):91–97. DOI: 10.1016/j.vetpar.2012.10.011

[125] Musa AM, Noazin S, Khalil EA, Modabber F. Immunological stimulation for the treatment of leishmaniasis: a modality worthy of serious consideration. Trans R Soc Trop Med Hyg. 2010; 104(1):1–2. DOI: 10.1016/j.trstmh.2009.07.026

[126] Miret J, Nascimento E, Sampaio W, França JC, Fujiwara RT, Vale A, Dias ES, Vieira E, Costa RT, Mayrink W, Neto AC, Reed S. Evaluation of an immunochemotherapeutic protocol constituted of N-methyl meglumine antimoniate (Glucantime®) and the recombinant Leish-110f® + MPL-SE® vaccine to treat canine visceral leishmaniasis. Vaccine. 2008; 26:1585–1594. DOI: 10.1016/j.vaccine.2008.01.026

[127] Borja-Cabréra GP, Santos FN, Santos FB, Trivellato FAA, Kawasaki JK, Costa AC, Castro T, Nogueira FS, Moreira MAB, Luvizotto MCR, Palatnik M, Palatnik-de-Sousa CB. Immunotherapy with the saponin enriched-Leishmune® vaccine versus immunochemotherapy in dogs with natural canine visceral leishmaniasis. Vaccine. 2010; (8):597–603. DOI: 10.1016/j.vaccine.2009.09.071

[128] Joshi J, Malla N, Kaur S. A comparative evaluation of efficacy of chemotherapy, immunotherapy and immunochemotherapy in visceral leishmaniasis an experimental study. Parasitol Int. 2014; 63(4):612–620. DOI: 10.1016/j.parint.2014.04.002

[129] Grandoni L. Canine leishmaniavaccines: still a long way to go. Vet Parasitol. 2015; 208(1-2):94–100. DOI: 10.1016/j.vetpar.2015.01.003.

[130] Palatnik-De-Sousa, C.B. Vaccines for canine leishmaniasis. Front Immunol. 2012;3:69. DOI: 10.3389/fimmu.2012.00069

[131] da Silva VO, Borja-Cabrera GP, Correia Pontes NN, de Souza EP, Luz KG, Palatnik M, Palatnik de Sousa CB. A phase III trial of efficacy of the FML-vaccine against canine kala-azar in an endemic area of Brazil (São Gonçalo do Amarante, RN). Vaccine. 2001; 19(9–10):1082–1092.DOI: 10.1016/S0264-410X(00)00339-X

[132] Costa-Pereira, C, Moreira ML, Soares RP, Marteleto BH, Ribeiro VM, França-Dias MH, Cardoso LM, Viana KF, Giunchetti RC, Martins-Filho OA, Araújo MS. One-year timeline kinetics of cytokine-mediated cellular immunity in dogs vaccinated against visceral leishmaniasis. BCM Vet Res. 2015;11:92. DOI: 10.1186/s12917-015-0397-6

[133] Borja-Cabréra GP, Correia Pontes NN, da Silva VO, Paraguai de Souza E, Santos WR, Gomes EM, Luz KG, Palatnik M, Palatnik-de-Sousa CB. Long lasting protection against canine kala-azar using the FML-QuilAsaponin vaccine in an endemic area of Brazil (São Gonçalo do Amarante, RN). Vaccine. 2002; 20(27–28):3277–3284. DOI: 10.1016/S0264-410X(02)00294-3

[134] Fernandes AP, Costa MM, Coelho EA, Michalick MS, de Freitas E, Melo MN, Luiz Tafuri W, Resende DM, Hermont V, Abrantes CF, Gazzinelli RT. Protective immunity against challenge with *Leishmania (Leishmania) chagasi* in beagle dogs vaccinated with recombinant A2 protein. Vaccine. 2008; 26(46):5888–5895. DOI: 10.1016/j.vaccine.2008.05.095

[135] Testasicca MCS, dos Santos MS, Machado LM, Serufo AV, Doro D, Avelar D, Tibúrcio AM, Abrantes Cde F, Machado-Coelho GL, Grimaldi G Jr, Gazzinelli RT, Fernandes AP. Antibody responses induced by Leish-Tec®, an A2-based vaccine for visceral leishmaniasis, in a heterogeneous canine population. Vet Parasitol. 2014;4(3–4):169–176. DOI: 10.1016/j.vetpar.2014.04.025

[136] Brasil. Ministério da Saúde. Esclarecimento sobre substituição do protocolo diagnóstico da leihsmaniose visceral canina. 2011. Nota técnica conjunta nr. 01/2011 -CGDT-CGLAB/DEVIT/SVS/MS.

[137] Grimaldi G Jr, Teva A, Ferreira AL, dos Santos CB, Pinto Id, de-Azevedo CT, Falqueto A. Evaluation of a novel chromatographic immunoassay based on Dual-Path Platform technology (DPP® CVL rapid test) for the serodiagnosis of canine visceral leishmaniasis. Trans R Soc Trop Med Hyg. 2012; 106(1):54–59. DOI: 10.1016/j.trstmh.2011.10.001

[138] Laurenti MD, de Santana Leandro MV Jr, Tomokane TY, De Lucca HR, Aschar M, Souza CS, Silva RM, Marcondes M, da Matta VL. Comparative evaluation of the DPP(®) CVL

rapid test for canine serodiagnosis in area of visceral leishmaniasis. VetParasitol. 2014; 205(3–4):444–450. DOI: 10.1016/j.vetpar.2014.09.002

[139] Schubach EY, Figueiredo FB, Romero GA. Accuracy and reproducibility of a rapid chromatographic immunoassay for the diagnosis of canine visceral leishmaniasis in Brazil. Trans R Soc Trop Med Hyg. 2014; 108(9):568–574.DOI: 10.1093/trstmh/tru109

[140] Peixoto HM, de Oliveira MR, Romero GA. Serological diagnosis of canine visceral leishmaniasis in Brazil: systematic review and meta-analysis. Trop Med Int Health. 2015; 20(3):334–352. DOI: 10.1111/tmi.12429

[141] Souza Filho JA, Barbosa JR, Figueiredo FB, Mendes AA Jr, Silva SR, Coelho GL, Marcelino AP. Performance of Alere™ immunochromathographic test for the diagnosis of canine visceral leishmaniasis. Vet Parasitol. 2016; 225:114–116. DOI: 10.1016/j.vetpar.2016.06.011

[142] de Oliveira IQ, Silva RA, Sucupira MV, da Silva ED, Reis AB, Grimaldi G Jr, Fraga DB, Veras PS. Multi-antigen print immunoassay (MAPIA)-based evaluation of novel recombinant *Leishmania infantum* antigens for the serodiagnosis of canine visceral leishmaniasis. Parasit Vectors. 2015; 8:45. DOI: 10.1186/s13071-015-0651-6

[143] Organização Panamericana de Saúde. WHO-PAHO Expert Consultation on Visceral Leishmaniasis in The Americas. Consulta de expertos OPS/OMA sobre leishmaniasis visceral em las Américas [Internet]. 2005. Available from: http://bvs1.panaftosa.org.br/local/File/textoc/LEANES_Inf_final_leish_2005.pdf#page=28 [Accessed: 2016-09-01]

[144] Brasil. Ministério da Saúde. Funasa: 20 years in the heart of Brazil. Brasília: Ministério da Saúde, Fundação Nacional de Saúde, 2011. 52 p.

[145] Zuben APBV, Donalísio MR. Difficulties in implementing the guidelines of the Brazilian Visceral Leishmaniasis Control Program in large cities. Cad Saúde Pública. 2016; 2(6): 1–11. DOI: 10.1590/0102-311X00087415

Chapter 3

Infectious Causes of Abortion, Stillbirth and Neonatal Death in Bitches

João Marcelo Azevedo de Paula Antunes,
Débora Alves de Carvalho Freire,
Ilanna Vanessa Pristo de Medeiros Oliveira,
Gabriela Hémylin Ferreira Moura,
Larissa de Castro Demoner and
Heider Irinaldo Pereira Ferreira

Additional information is available at the end of the chapter

http://dx.doi.org/10.5772/65330

Abstract

Problems in gestational development in dogs can be determined by infectious and non-infectious causes. Among the non-infectious causes, trauma during pregnancy, genetic characteristics of the animal, deficit nutrition, thyroid dysfunction, maternal problems and hormonal disorders are found. The majority of the cases are in relation to infectious diseases, one should consider viral, bacterial, fungal and protozoal, which can interfere directly or indirectly in the foetal development. The progression of foetal development may be affected by the direct action of the microorganisms to overcome the placenta, but they are also able to affect pregnancy and release placental toxins by inflammatory processes and, may still cause maternal pathologies, which entail problems such as hyperthermia, hypoxia and endotoxemia, which can result in abortion. Several diseases can trigger pregnancy loss in dogs. This action can be direct by microorganisms, as well as indirectly triggering other problems that lead to abortion. This chapter discusses the infectious aetiologies of reproductive failures (abortion, stillbirth and neonatal death) in bitches.

Keywords: Puppies, Bitches, reproductive failure, infectious causes, diagnostic

1. Introduction

Problems in gestational development in dogs can be caused mainly by infectious diseases. The progression of foetal development may be affected by the direct action of microorganisms to overcome the placenta, but they are also able to affect pregnancy and release placental toxins by inflammatory processes and may still cause maternal pathologies, which entail problems such as hyperthermia, hypoxia and endotoxemia, which can result in reproductive failures (abortion, stillbirth and neonatal death) [1].

2. Bacterial diseases

2.1. Brucellosis

The bacterium *Brucella canis* (small, aerobic Gram-negative coccobacilli, which stain red using the modified Ziehl-Neelsen technique) stands out as one of the main bacterial causes of pregnancy loss in bitches [2]. The species *B. abortus, B. melitensis* and *B. suis* have also been found in dogs, where it is believed that the natural infection occurs after the ingestion of placenta and aborted foetuses. The main source of infection is through vaginal and seminal secretions from infected animals, although bacteria are shed in faeces, milk, saliva, and nasal and ocular secretions [2]. The *B. canis* may be present for a long time in dogs without exhibiting clinical signs. After the initial exposure, the bacteria reach the bloodstream in about three weeks. Subsequently, the pathogen can infect the genital tissues enabling a continuous release of the agent, which may be recurring for months or even years. In turn, canine brucellosis can result in infertility, difficulties in pregnancy, early embryonic death, foetal resorption and late abortion. The clinical signs associated with Brucellosis in dogs are not pathognomonic and due the lack of the lipopolysaccharide antigen associated with endotoxemia in bitches, it is rarely systemic ill and fever [3]. Clinical signs reflect the localization of the bacteria in extra reproductive tract sites such as the eye, intervertebral disc spaces, and reticuloendothelial system. Brucellosis causes spontaneous late abortion in a healthy bitch, most commonly occurs from days 30 to 57, accompanied by a vaginal discharge lasting up to 6 weeks. Earlier abortions can occur but may be incorrectly reported as conception failure since the bitch typically ingests aborted foetuses. Early embryonic death and foetal resorption can occur within 10 – 20 days post-mating. Many bitches that abort will subsequently have normal litters, although puppies born to infected bitches contain both live and dead pups, although most live pups die shortly thereafter. Aborted puppies usually appear partially autolysed, with lesions of generalized bacterial infection, including subcutaneous oedema and degenerative lesions in the liver, spleen, kidneys and intestines. Seroprevalence studies indicate that canine brucellosis is widespread in the Americas. Isolation (placenta, lymph nodes, prostate, and spleen are suitable samples for culture) and identification of *B. canis* is the gold standard, however, serology (the most accurate serological test currently available is the agar gel immunodiffusion test-AGID) and polymerase chain reaction (PCR) are widely used to diagnose canine brucellosis [3, 4]. *B. canis* is not considered a significant zoonosis and, serious illness can occur in immunocompromised patients by direct contact with infected animals or through occupational aerosol exposure, but the owners should be aware of the zoonotic potential of this disease [4].

2.2. Others bacterial agents and Rickettsias

The *Ehrlichia canis* and *Anaplasma platys* have been found in dogs attended in veterinary hospitals and clinics in various countries and states in Brazil [1]. Anaemia and thrombocytopenia are the main clinical signs found in dogs infected with these agents in Brazil. However, it is noted that these dogs more often develop anaemia and thrombocytopenia. So, clinical signs may vary according to geographical variations and pathogenicity. These agents are not yet known to cause abortion directly but have been reported in abortion in anaemic bitches infected. Results also indicate that animals with infectious anaemia may be more susceptible to suffer reproductive failures than animals with normal haematological values [1] (**Figure 1**).

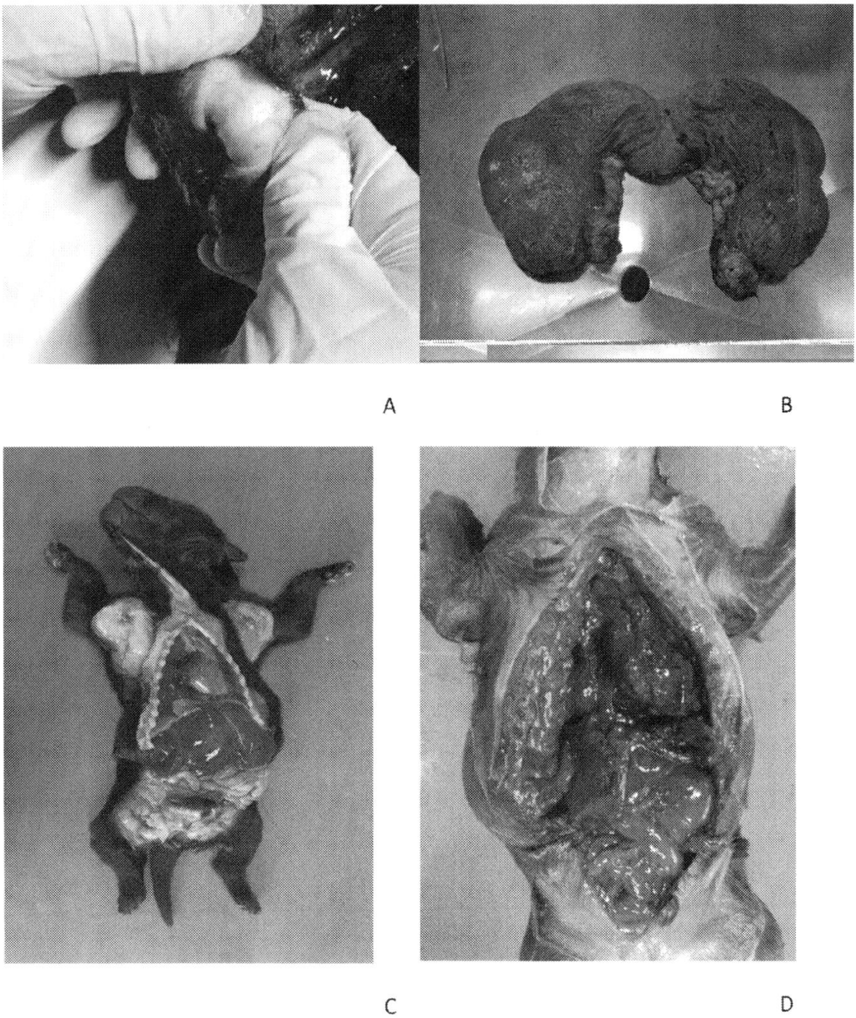

Figure 1. Macroscopic lesions caused after abortion due anaemia occasioned by vector-borne disease. (A). Anaemic vulva after intense parasitaemia by ticks; (B). uterus with haemorrhagic serous after episode of abortion; (C). hepatosplenomegaly in aborted foetus with 56 days of gestation; (D). foetus with blackened organs and in the autolysis process after stillbirth in late pregnancy.

Escherichia coli is the most common bacterium isolated from the canine vagina and is also commonly cultured from the uteri of bitches with metritis and pyometra. *E. coli* produces an endotoxin that may result in pregnancy loss in the bitch. *Streptococcus* spp are bacteria that are physically present in the skin and mucosa of dogs and cats. Some microorganisms belonging to this group have been related to the occurrence of neonatal sepsis, abortion and metritis [2, 3]. βH *Streptococcus* is known cause metritis, pyometra, placentitis and abortion, commonly linked to ascending infection. *Streptococcus* βH is a major cause of neonatal death. Typically, newborns are infected in the mother's birth canal, through the umbilical cord or, less commonly, from the udders with mastitis. Bitches have high bacterial load in the vaginal canal, which persist throughout pregnancy leading to the infection of the uterus and foetus [2, 3].

3. Protozoal diseases

Some protozoan species are capable of infecting dogs and can also infect humans, triggering zoonoses. The existence of protozoal coinfection in dogs are already established, one of the combination of *Toxoplasma gondii* and *Neospora caninum* sometimes increased by the presence of *Leishmania* spp. These types of infections exacerbate the clinical state of the animal by co-existence of various diseases [5].

The *T. gondii* causes toxoplasmosis and its definitive hosts are the Felidae. However, its prevalence in dogs is high. Transmission can occur vertically through congenital infection, however, horizontal transmission is the main mean of infection. Dogs are the definitive hosts of *N. caninum* that causes Neosporidiose disease, which can be transmitted horizontally or vertically, and it has maintained the parasite for generations amongst the canines [84].

3.1. Toxoplasmosis

Toxoplasma gondii, the protozoan that causes a coccidiosis in felines can infect mammals (including humans), birds and reptiles, mainly affecting the central nervous system and occasionally, the reproductive system, muscles and visceral organs [6]. This protozoan is an obligate intracellular parasite and its definitive hosts are the Felidae, which excrete oocysts that house sporozoites, an infectious stage of the parasite. As for the intermediate hosts and therefore not Felidae, which shelter tissue cysts, it was possible to observe two infectious stages of the organism: tachyzoites and bradyzoites [7]. The oocytes containing the sporozoites have a double-walled, spherical shape and resistance to environmental conditions, Being the parasitic manner that promotes their strength and dissemination. They are developed by a entero-epitelial cycle inside the felids through sporogony, and are then excreted in their immature form with the faeces of these animals. The tachyzoites Form occurs in the active proliferation stage of *T. gondii,* or the acute phase of infection Parasitizing preferably macrophages and monocytes but with the potential to infect any nucleated cell. This parasitic stage has an arc shape, is mobile and provides rapid multiplication by endodyogeny. In the chronic phase of infection are found the bradyzoites located in parasitophorous vacuole of a cell whose membrane forms the capsule of the tissue cyst. They have slow multiplication, also endodyogeny, intracisto, where it is protected from the action of the host immune system and drugs, allowing its existence for months and even years. They are poorly resistant to high temperatures [8].

T. gondii has worldwide distribution and is considered as the most cosmopolitan of all the causative parasite zoonoses. It is an opportunistic organism, independent of the host and has relevance related to animal production and public health by having transmitted through food from infected animals. In transmission between animals, the Felidae assume the lead role as a source of infection [8]. Thus, the transmission can occur by congenital way, via ingestion of infected tissue or contaminated food and water, in addition to transmission by breastfeeding, through transfusion of body fluids or by transplantation of tissues or organs [7].The entero-epitelial biological cycle is unique to the definitive hosts in which the oocysts are released in the faeces. Intermediate hosts are infected by ingestion of oocysts and cysts. Oocysts must sporulate in the environment to become infective and once sporulated and ingested by potential hosts, will promote asexual reproduction and infect the animal that ingested it, leading to the formation of tissue cysts. Cysts, in turn, may be ingested by host carnivores, infecting them. Ingestion of raw or undercooked meats are the main sources of this type of transmission. Congenital transmission and placenta are more common in women, sheep, goats and nuts. However, studies have shown that infected female dogs with 56, 40 and 32 days of gestation showed congenital infection that may have led to abortion and thus confirm this relationship.

The evidence of facts relating to such a manifestation of the parasite in dogs is scarce. What is known is that in its life cycle, the *T. gondii* reaches different host tissues, including male and female reproductive organs of intermediate hosts, which may cause certain adverse effects on the reproductive function. In animals, toxoplasmosis is associated with reproductive failure in males related with some evidence of venereal transmission of *T. gondii* [9]. Researchers have shown that four bitches inseminated with canine semen samples containing 1×10^6 tachyzoites of *T. gondii* RH strain, had embryonic resorption in two of them and the protozoan was found in the brain of the four offsprings [10]. The results of this study show, among other things, that dogs with infected semen can cause reproductive problems and, consequently, the vertical transmission to offspring. Additionally, transmission of tachyzoites to the milk intake neonates (via lactogenic) shortly after birth [11] has been reported. The severity and type of clinical signs presented by the Animals infected with *T. gondii* are variable depending on the degree of tissue injury and location. There is no established reason regarding the different presentations of clinical signs and their absence in dogs. It is assumed that all types of cells are susceptible and some of these differences can be assigned to factors such as age, sex, host species, *T. gondii* strain, body number and stage of ingested parasites [8]. Specifically in dogs, the occurrence of clinical signals from toxoplasmosis is associated with concomitant infections such as distemper, ehrlichiosis, neosporidiose and leishmaniasis.

Animals infected with *T. gondii* usually show no clinical symptoms and undergo the primary infection for a latent infection or a chronic stage. In dogs, infection with this parasite is very common, being demonstrated by several studies of serological prevalence. Already the clinical form of infection is rather sparse, usually presenting itself in young animals, usually associated with immunosuppressive factors, infections or to canine distemper virus. Canine symptoms can concentrate in the gastrointestinal tract and respiratory systems and in neuromuscular or be from generalized infection. Therefore, clinical signs usually identified in dogs with toxoplasmosis are ataxia, diarrhoea and respiratory disease and focal necrosis in areas of the lung, liver and brain may also occur, triggering numerous clinical signs [12]. Therefore, toxoplasmosis in dogs and other animals can show symptoms quite similar to many infectious diseases, the Neosporidiose one with the greatest similarity.

The possible interference of Toxoplasmosis in the reproduction of dogs as a whole has been studied in 1970 and found out that toxoplasma infection in dogs at various gestational stages can cause mortality of the puppies from the 4th to 75th postnatal days [13]. Still, abortion and foetal death has been observed, in the middle third and final pregnancy in dogs experimentally infected with *T. gondii* orally (oocysts) and subcutaneously (tachyzoites) [14]. Later, reproductive problems related to canine toxoplasmosis were again observed, when naturally infected dogs with 30 days of gestation were artificially reinfected. The puppies from infected mothers were positive serologically at birth, but without symptoms, except a neonate who was weakened. With additional tests, the parasite was detected in saliva samples, milk and urine of these animals still being detected positive immunostaining of cysts and / or tachyzoites by immunohistochemistry in 23 organs of experimental animals [15, 16]. Soon, placental transmission and subsequent foetal infection with *T. gondii* in cats and dogs, have been experimentally observed and their association with foetal death have been found.

Diagnosis of toxoplasmosis is mainly based on clinical suspicion since the symptoms can be similar to other infectious diseases. Therefore, attention should be paid to the epidemiological information and data collected during the interview, making a request for additional tests, a prerequisite. In dogs, the presence of apathy, rhinorrhea, conjunctivitis, pneumonia, fever, convulsions, paralysis, diarrhoea and lymphadenopathy may be clinical signs of *T. gondii* infection, requiring additional tests to confirm the diseases [6]. The identification of *T. gondii* presence can be accomplished by in vitro and in vivo isolation, serology, histopathology, immunohistochemistry and PCR. In general, the serological diagnosis is the most used, mainly through modified agglutination test (MAD) and indirect immunofluorescence (IFA), which is the preferred test [17]. Whichever method is used, it is important to collect samples at intervals of about two weeks to determine seroconversion, indicating recent infection [18]. When there is abortion, should be referred to the placenta and the foetus (paying attention to the foetal brain), cooling temperature, as soon as possible to enable the laboratory diagnosis for toxoplasmosis. Submissions will be used for histopathology associated or not with immunohistochemical tissues with compatible lesions [8].

In cases of acute systemic infection in dogs changes can be observed in haematological parameters, such as nonregenerative anaemia, neutrophilic leucocytosis, lymphocytosis, monocytosis and eosinophilia. The biochemical changes are consistent with increased serum activities of alanine aminotransferase (ALT) and alkaline phosphatase may occur when there is liver necrosis. If necrosis is acute, bilirubin levels are likely to be increased. In the event of muscle necrosis, serum creatine kinase activity is also increased [7]. The establishment of the diagnosis can be made through serological tests, but there is no absolute serologic test that can definitely confirm the diagnosis of toxoplasmosis. Furthermore, only the detection of antibody against the parasite is not sufficient for the establishment of serological diagnosis since previously infected dogs may also exhibit antibodies response. Some tests that may be ordered in addition to the IFT and MAD, such as the reaction of Sabin-Feldman (SF), the enzyme linked immunosorbent assay (ELISA), complement fixation test (CF) and the reaction of indirect haemagglutination (HI) [8].

The IFA identifies antibodies through specific fluorescent conjugates of IgG and IgM. The presence of IgG is related to a previous exposure, suggesting active and recent infection. The MAD evaluates IgM antibodies indirectly by subtracting antibody titres present in treated and untreated sera [19]. The SF reaction has a high sensitivity and specificity and allows the antibody titration within a few days post-infection. The enzyme linked immunosorbent assay allows detection of IgM response. In the CF test, different parasite antigens are used and that can be identified as testing positive in case of an early infection. In short, the interpretation of the results: a high titre of IgM with low or zero IgG indicates a disease evolving, otherwise, the high quantitation of IgG and low or zero IgM indicate a chronic state of the disease. Whatever the outcome, a new serology test should be done after 15–21 days due to the possibility of severe immunosuppression until the second post-infection week, which may lead to false-negative results [8]. Identifying the *T. gondii* in tissue or bodily fluids is possible by polymerase chain reaction that detects the protozoa in biological samples. However, this detection which is based on PCR is still quite limited, as it requires special equipment, which are not available in any clinical laboratory. In addition, the realization of body isolation requires relatively vast and considerable experience time. However, care should be taken that the molecular detection of the parasite does not replace the serological methods in the diagnosis of the disease, making it imperative to associate the two diagnostic methods to identify the presence of infection and the determination of the disease stage [8].

Various infectious and non-infectious diseases should be evaluated in the differential diagnosis of toxoplasmosis. In dogs, the most relevant are distemper, neosporosis, isosporiasis and strongyloidiasis [8]. In aborted foetuses, macroscopically it is possible to observe necrotic foci white-greyish punctate located in the lungs and liver, as well as pulmonary congestion and heart pallor. The central nervous system (CNS) is marked with congestion of the brain and cerebellum [7].

Treatment of toxoplasmosis is based on suppression of the replication agent when the disease presents itself in the acute form, and emphasizes the importance of early diagnosis and consistent implementation of measures leading to the reduction in disease transmission [20]. Clindamycin is the drug of choice for the treatment of dogs and cats. The drug administration once started, the clinical signs tend to initiate regression after 24 – 48 hours. In addition to clindamycin, the combined use of pyrimethamine and fast-acting sulphonamides (e.g., sulfadiazine, sulfamethazine and sulfamerazin) is valid in the treatment of systemic toxoplasmosis infection [7].

Prevention should be taken to avoid ingestion and contact with oocysts and cysts. So for pets that eat meat, the meat should be well cooked. As a complement to prevention, measures should be taken to avoid these same animals hunt and/or eating mechanical hosts such as cockroaches, flies, worms and rodents, potential or intermediate hosts. The prevention becomes even more important because there is no vaccine available for humans or animals [8]. General care in food hygiene, whether in urban or rural areas, considering proper cooking of meat and milk boiling, are the main preventive measures of toxoplasma infection to humans. Personal hygiene and comprehensive hand washing performed after handling raw meat, for example, assists in preventing the disease. Generally, prevention of human toxoplasmosis is based on the maximum avoidance of exposure to susceptible hosts. This disease is not notifiable, except for outbreaks [7].

3.2. Neosporosis

Neospora is a protozoan belonging to the phylum Apicomplexa, Sporozoa class, Eucoccidiida order and Sarcocystidae family [21]. *Neospora caninum* is a parasite coccidia, intracellular exclusively, forming cysts that causes neosporidiose, a disease that has been disseminated in the continents generating a high rate of infected cattle and dogs [22]. In the genus *Neospora* were identified only two species *N. caninum* [23] isolated from dog brain and *N. hughesi* [24, 25] isolated from the brain and spinal cord of horses [24]. It has not been described any cases of the disease in humans [26]. Dogs (*Canis lupus familiaris*) [27], coyotes (*Canis latrans*) [28], dingo (*Canis lupus dingo*) [29] and gray wolves (*Canis lupus*) [30] have been identified as definitive hosts of the parasite and are significant in the transmission of the parasite to other animals. The prevalence of antibodies against *N. caninum* in dogs is the result of research in various parts of world, which has reached from 4 to 54.2% [31, 32]; this percentage can be varying because of habitat, age, living of dogs with cattle, diet and serological technique employed etc., among others [33]. The life cycle of *N. caninum* has three forms: tachyzoites, bradyzoites in the cysts and sporozoites in the oocysts. However, the tachyzoites and bradyzoites are intracellular stages identified in intermediate hosts [34]. The tachyzoites are responsible for the acute phase and has the ovoid or circular shape. Bradyzoites are in latent stages, through tissue cysts that are resistant to acid solution pepsin [35].

Horizontal transmission (postnatal) is due to the ingestion of water or food contaminated with oocysts eliminated by dogs, especially in cases of abortion outbreaks. It is identified as the association between seroprevalence and abortions in bovines, when the presence of dogs could increase the incidence of the disease in both species. Dogs present in farms were identified with greater prevalence of infection than in urban area [36]. This is due to the dogs ingest infected bovine foods such as foetuses, foetal membranes and fluids [37]. In dogs, neosporidiose can develop neuromuscular, cardiac, pulmonary and skin lesions changes, there still descriptions of dermatitis, cardiomyosite and pneumonia in this species [38, 39]. Vertical transmission via lactogenic have been reported in calves experimentally and the presence of DNA of *N. caninum* was also observed in the colostrum of cows infected, reporting the possibility of transmission of *N. caninum* by colostrum [40].

Contamination can occur vertically when a bitch with subclinical infection transmits *N. caninum* to its foetuses and litters can be born infected [41]. The congenitally infected animals are the most severe cases; however, the infection can occur in animals of all ages. Young animals manifest hind limb paralysis that occurs at an accelerated rate, and incoordination can cause paresis of hind limbs in adult dogs [42, 43]. For a reliable diagnosis: detailed history and description, good physical examination and laboratory tests are necessary [42]. Haematological and biochemical tests are not enough to confirm the diagnosis of this disease, since the laboratory diagnosis is affected by means of serological and parasitological examinations. However, the muscle biopsy is considered important for the diagnosis of neosporidiose [23, 44]. The use of immunohistochemistry is also of paramount importance to identify tachyzoites and cysts in the tissues [37]. Analysis of the cerebrospinal fluid is another form of diagnosis that can facilitate the identification of this disease. The inflammatory cell count demonstrates an inflammatory or infectious situation and tachyzoites display may indicate protozoal encephalomyelitis, and can also identify bradyzoites in cerebrospinal fluid. In the dog faeces, the oocysts of *N. caninum* can be observed using the flotation technique [37, 45].

The PCR is essential to detect the DNA of the parasite in the faeces of the definitive host and thus used to confirm the diagnosis. This technique is also used to perform DNA detection in the intermediate host [46, 47]. Serological tests such as indirect immunofluorescence assay, agglutination test Neospora (NAT) and multiple tests of ELISA were made for diagnosis in dogs [48]. IFA was the initial test to identify antibodies against *N. caninum*. Immunofluorescence with greater than or equal to title 1:50 demonstrates dog contact with the agent. The NAT test demonstrates the agglutination of tachyzoites in the presence of specific antibodies contained in the serum, abolishing the use of secondary antibodies used in the previous tests. This test has identified specificity and similar sensitivity of the IFA [21, 49].

Macroscopic lesions caused by *N. caninum* is rare, however in experimental infections, it was identified necrosis of the foetal placental villi, necrosis and inflammation [50]. For the treatment of dogs clindamycin (11-22 mg/kg, BID, TID), sulphonamides (15 mg/kg bid) and pyrimethamine (1mg/kg SID) [51] are used. However, the treatment efficiency is not effective, but there are data that demonstrate an effective response against neosporosis in the symptoms of adult dog administration associated with pyrimethamine and sulphadoxine for 1 month [52]. For the prophylaxis and control of *N. caninum* infection, the reproduction of positive dogs that have already demonstrated compatible symptoms or have calved infected puppies and sick should be prevented. It should also be able to prevent these carnivores eat foods such as meat or raw entrails, and especially farm animals [42]. To prevent the ingestion of aborted foetuses and placental membranes should discard these materials in an appropriate place, since we do not have efficient commercial vaccines on the market [53].

4. Viral diseases

4.1. Canine herpesvirus

The canine herpesvirus 1 (CHV-1) has a worldwide distribution and is associated with respiratory and reproductive diseases in dogs [54]. The CHV-1 was isolated in several countries, with a disease considered enzootic for dogs [55]. The first study to report this agent associated with fatal haemorrhagic disease in puppies was Carmichael et al. in 1965 [56]. The *Canid alphaherpesvirus* 1 species refers to family *Herpesviridae*, subfamily *Alphaherpesvirinae* and genus *Varicellovirus* [57]. This virus consists of double-stranded DNA, has icosahedral infecting only dogs or cells canine origin. This specificity is due to the presence of specific receptors on the cell surface, such as the glycoprotein D (haemagglutinin) [54, 58]. As for CHV-1 characteristics of the environment, there is a higher incidence and spread of the virus in kennels than in the home environment. The presence of this agent in serum from canis can get up to 100%. This fact is mainly due to agglomeration and poor hygiene conditions in places [58]. Moreover, the disease manifests itself in seasonal way, accentuating in cold weather because the virus is unstable and sensitive to higher temperatures [58, 59].

Among the infectious diseases of viral origin, the CHV-1 stands out as one of the main viral cause of abortion and neonatal mortality in dogs [60]. The infection caused by this virus during pregnancy can lead to abortion, stillbirth, embryonic resorption, premature birth and neonatal death [61, 62]. It can result in infertility, birth of mummified foetuses, weak puppies, or premature sick [55]. Its horizontal transmission can occur due to direct or indirect contact between dogs. This contact can happen through nasal secretions, semen and contaminated aerosols, regardless of sex or age distinction, being observed an increased susceptibility in puppies less than 2 months [63]. Animals without updated immunization record, with vaccination failures and maternal immunization also become more prone to infection [64]. Vertical transmission occurs from mother to foetus through the placenta [55]. In some cases the infection can reach the uterus resulting in foetal death and still birth of the offspring [65]. After infection of the cell by the virus in the cell nucleus will be synthesis of viral DNA and nucleocapsids. As the viral envelope from the nuclear membrane. The virus then travels through the endoplasmic reticulum and Golgi, and subsequently released to infect new cells [58].

Infected adult dogs often do not show apparent symptoms. In them, the infection is often subclinical. However, in newborns and puppies with 1–2 weeks of life may develop systemic disease that may result in a generalized necrotizing haemorrhagic disease [55, 56]. Still, this pathogen has an important characteristic of latent infection remaining in a state of latency in lymph nodes and lymphoid tissues of the oronasal mucosa and genital [58, 66]. So that makes the diagnosis difficult due to the absence of clinical signs. In this condition, the presence of factors such as pregnancy, stress, immunity reduction, diseases and use of corticosteroids can reactivate the virus [58], with the possible occurrence of necrosis in the placenta in pregnant infected bitches [65]. In females, papulovesicular in genitalia and oral lesions can be visualized, and in males, may be similar lesions on the penis and release the virus by semen [58]. The histopathologic examination of the liver, lung and kidney in adult dogs reveals haemorrhages with necrotic foci and intra-nuclear inclusion bodies. Puppies contaminated by CHV-1 usually die resulting from systemic disease. The consequences of pathogen infection during pregnancy in bitches will depend on the stage of gestation when infection occurred. Histopathological and post-mortem exams may be observed mummified and calcified foetuses, foetuses with progressive multifocal haemorrhagic necrosis in various organs, and haemorrhagic foci in uteri and placentas. An increased virus concentration has also been observed in the adrenal glands, kidneys, lungs, spleen and liver. In the post-mortem examination the presence of serous fluid and haemorrhage in the pleural and peritoneal cavities is observed. And especially in canine puppies infected by the oronasal route may have meningoencefatite after birth [58].

Diagnosis can be accomplished through fluids and vaginal swabs collected and nasal secretions, or using tissues from foetal organs or adult dogs after necropsy. The antigen or genetic material in aborted foetuses or newborns can be extracted from humours, liver, adrenal glands, lungs, spleen, kidneys and lymph nodes. With the collection of the appropriate material, histopathological tests, biochemical, immunofluorescence microscopy and molecular tests can be performed. The CHV-1 presence can be confirmed in samples by polymerase chain reaction [63] and by the sequencing, comparing similarity between the surrounding sequences. PCR allows for viral detection even in animals that have the latent infection [58]. Obtaining the history and medical records of these animals is also essential for a complete and accurate diagnosis [67]. The treatment of this disease in puppies is difficult due to the rapid development of infection and mortality that occur before the diagnosis is established [58]. However, despite this agent being seen as the main cause of abortion and foetal mortality in dogs, other pathogens may be associated directly or indirectly with these consequences, such as loss of pregnancy by systemic infectious anaemia [1]. Vaccination and appropriate sanitary measures are essential to prevent viral spread among animals in kennels [63].

4.2. Canine minute virus

Another viral agent that can cause severe disease in newborns, transplacental infections and embryonic resorption in dogs is *Canine minute virus* (CnMV) or CPV-1 (*Canine Parvovirus*-1) [68]. This agent sets off a disease currently considered endemic [69]. The consequences of infection with this virus in pregnant females may vary according to the time of infection during pregnancy and may cause embryonic resorption, stillbirth, neonatal mortality and abortion [65, 70], or initiate respiratory, cardiac and enteritis problems in puppies [65]. The CnMV is a parvovirus that belongs to *Parvoviridae* family, subfamily *Parvovirinae* and genus *Bocavirus* [57]. It is a very small virus approximately 22 nm in diameter, single-stranded DNA and has no viral envelope. This virus has tropism for lymphoid and embryonic tissues, bone marrow, myocardium and intestinal epithelium [69].

The clinical signs found in CnMV infection may vary, be unapparent, or apparent as respiratory problems, enteric disease and reproductive disorders [71]. Infected puppies less than 1 month of age may have mild symptoms or accelerated death, depression, lack of appetite, acute myocarditis, respiratory failure and enteritis. The viral action mechanism contributes to reducing phagocytosis by monocytes promoting immunosuppression [68, 72]. The transplacental infections can cause disease in subclinical phase, deformations of foetuses and the loss of pregnancy. The consequences in pregnant bitches may vary according to the time of infection. In the first half of the pregnancy, stillbirth and embryo resorption process can occur after infection. In the second half, observed a larger number of stillborn and weak puppies [71].

As for post-mortem examinations on puppies, pneumonia, enteritis, myocarditis, oedema and atrophy of the thymus have been observed. In relation to the histopathologic findings, viral presence in the epithelial cells of the intestinal crypts and cardiomyocytes are observed. Other changes found are hyperplasia of the interstitial crypts, myocardial necrosis, pneumonia, and depletion of lymphocytes in the thymus and other lymphoid tissues [71]. The samples for diagnosis may vary mostly from foetal or neonatal tissues of the myocardium, intestine, lungs, kidneys and faeces. The diagnosis of the CnMV can be accomplished by direct methods such as viral isolation, immunohistochemistry, electron microscopy, direct ELISA, conventional PCR and RT-PCR, of the tissue, and/or faeces and/or enteric contents. For detection of intranuclear inclusion bodies, haematoxylin-eosin staining or immunofluorescence using specific antibodies [73] can be used. As for indirect methods of serological diagnosis may used the indirect ELISA and haemagglutination inhibition, allowing the study of the prevalence of disease [74].

The treatment of parvovirus is based on the recovery of electrolyte balance and in the prevention of secondary infections using antibiotics. Attenuated vaccines of CnMV in dogs provide superior immunity than inactivated, and are safer. In newborn puppies, it is suggested the warming of them and to maintain nutrition and adequate hydration [74].

4.3. Other viral diseases

Other viral infections known to cause abortions, stillbirths and neonatal death in bitches are Bluetongue (BTV), canine distemper and canine adenovirus-1 [75]. The Bluetongue is a disease transmitted by arthropods especially in ruminants. Infection of dogs is currently thought to be by oral ingestion of infected meat or meat products rather than through vector feeding [76]. There is evidence of direct transmission of the agent to the dog. Abortion and stillbirth are consequences of infection of this agent in pregnant bitches [77]. Direct transmission and differences in canine susceptibility to certain serotypes of the virus are not well elucidated in dogs [78]. The canine distemper virus (CDV) may also affect gestation by the weakness of maternal health inducing abortion, and in rare cases can cross the placenta and lead to abortion or foetal infection [79, 80]. The abortion can originate from a systemic infection in dogs or transplacental infection [81]. The infection by this virus can still result in stillbirth and congenital infections in dogs. Transplacentally infected puppies can develop neurologic signs within 6 weeks after birth [79, 82]. The canine adenovirus-1 (CAV-1) may be associated with fatal pneumonia in pups less than 1 month of age [83]. However, this virus can result in miscarriage, with or without foetal infection. Abortion can be a result of stress caused by the disease [79, 80].

5. Final considerations

Infectious causes are still the most responsible for reproductive failure in dogs through the direct action of the pathogen in the foetus and in placenta; however, we must always try to reduce the chances of reproductive failures that occur due to systemic action of the infectious agent in the mother through the early diagnostic and treatment.

Author details

João Marcelo Azevedo de Paula Antunes*, Débora Alves de Carvalho Freire, Ilanna Vanessa Pristo de Medeiros Oliveira, Gabriela Hémylin Ferreira Moura, Larissa de Castro Demoner and Heider Irinaldo Pereira Ferreira

*Address all correspondence to: joao.antunes@ufersa.edu.br

Universidade Federal Rural do Semi-Árido–UFERSA, and veterinarian at Veterinary Hospital Jerônimo *Dix-Huit* Rosado Maia, Mossoró, RN, Brazil.

References

[1] Freire DAC. Abortion and fetal death in bitches with infectious anemia (Dissertation). Brazil: Universidade Federal Rural do Semi-Árido; 2016. 49 p.

[2] Givens MD, Marley MSD. Infectious causes of embryonic and fetal mortality. Theriogenology. 2008;**70**:270–285.

[3] Graham EM, Taylor DJ. Bacterial reproductive pathogens of cats and dogs. Veterinary Clinics Small Animals. 2012;**42**:561–582.

[4] Greene CE, Carmichael LE. Canine brucellosis. In: Greene CE, editor. Infectious Diseases of the Dog and Cat. WB Saunders Co. Rio de Janeiro, Brail; 2006. pp. 369–381.

[5] Pretzer SD. Bacterial and protozoal causes of pregnancy loss in the bitch and queen. Theriogenology. 2008;**70**:320–326.

[6] Koch MO. Isolamento In vitro isolation of Neospora caninum and Toxoplasma gondii in semen of naturally infected dogs. Curitiba: Universidade Federal do Paraná; 2014.

[7] Dubey JP, Lappin MR. Toxoplasmosis and neosporosis. In: EC Greene, editor. Infectious diseases of dogs and cats. 4th ed. Rio de Janeiro: Guanabara Koogan; 2015. pp. 842–864. ISBN: 9788527726900.

[8] da Silva RC, da Silva AV. Toxoplasmosis in domestic animals. In: Megid J, Ribeiro MG, Paes AC, editors. infectious diseases in production and companion animals. 1st ed. Rio de Janeiro: Roca; 2016. pp. 1040–1053. ISBN: 9788527727891.

[9] Dalimi A, Abdoli A. *Toxoplasma gondii* and male reproduction impairment: A new aspect of toxoplasmosis research. Jundishapur Journal of Microbiology. 2013;**6(8)**:e7184. DOI: 10.5812/jjm.7184.

[10] Arantes TP, Lopes WDZ, Ferreira RM, Pieroni JSP, Pinto VMR, Sakamoto CA, Costa AJ. *Toxoplasma gondii*: Evidence for the transmission by semen in dogs. Experimental Parasitology. 2009;**123**:190–194. DOI: 10.1016/j.exppara.2009.07.003.

[11] Dubey JP. *Toxoplasma, Neospora, Sarcocystis*, and other tissue cyst-forming coccidia of humans and animals. In: Krier JP, editor. Parasitic Protozoa. 6th ed. San Diego: Academic; 87; 1993.

[12] Dubey JP. Toxoplasmosis in dogs. Canine Practice. 1985;**12**:7–28. ISSN: 1057–6622.

[13] Helley DM. Toxoplasmosis. Journal of Small Animal Practice. 1970;**10**:627–629. ISSN: 1748–5827.

[14] Bresciani KDS, Costa AJ, Toniollo GH, Sabatini GA, Moraes FR, Paulillo AC, Ferraudo AS. Experimental toxoplasmosis in pregnant bitches. Veterinary Parasitology. 1999;**86 (2)**:143–145. DOI: 10.1016/S0304-4017(99)00136-3.

[15] Bresciani KDS. Study of Toxoplasma gondii infection in naturally infected pregnant bitches. Jaboticabal: Universidade Estadual Paulista; 2003.

[16] Bresciani KDS, Costa AJ, Toniollo GH, Kanamura CT, Luvizzoto MCR, Morais FR, Perri SHV. Transplacental transmission of Toxoplasma gondii in infected pregnant dogs. Revista Brasileira de Parasitologia Veterinária. 2004;**13(1)**:214. ISSN: 0103-846X.

[17] Camossi LG, Silva AV, Langoni H. Serological survey of toxoplasmosis in horses in the Botucatu region. Arquivo Brasileiro de Medicina Veterinária e Zootecnia. 2010;**62(2)**:484–488. ISSN: 1678–4162.

[18] Hill DE, Dubey JP. *Toxoplasma gondii*: Transmission, diagnosis and prevention. Clinical Microbiology and Infection. 2002;**8**:634–640. DOI: 10.1046/j.1469-0691.2002.00485.x.

[19] Silva FWS, Alves ND, Amóra SSA, Teixeira FHV, Accioloy MP, Carvalho CG, Nóbrega RM, Filgueira KD, Feijó FMC. Toxoplasmosis: Review. Ciência Animal. 2006;**16(2)**:71–77. ISSN: 1809–6891.

[20] Xue J, Jiang W, Chen Y, Liu Y, Zhang H, Xiao Y, Qiao Y, Huang K, Wang Q. Twenty-six circulating antigens and two novel diagnostic candidate molecules identified in the serum of canines with experimental acute toxoplasmosis. Parasites & Vectors. 2016;**9 (1)**:374. DOI: 10.1186/s13071-016-1643-x.

[21] Dubey JP, Beattie CP. Toxoplasmosis of Animals and Man. Boca Raton: CRC; 1988.

[22] Dubey JP, Schares G, Ortega-Mora LM. Epidemiology and control of neosporosis and *Neospora caninum*. Clinical Microbiology Reviews. 2007;**20**:323–367. DOI: 10.1128/CMR.00031-06.

[23] Dubey JP, Lappin MR. Toxoplasmosis and neosporosis. In: Greene CE, editor. Infectious Diseases of the Dog and Cat. 2nd ed. Philadelphia: WB Saunders Company, Cap. 90;1988. pp. 493–503.

[24] Marsh AE, Barr BC, Packham AE, Conrad PA. Description of a new Neospora species (Protozoa: Apicomplexa: Sarcocystidae). The Journal of Parasitology. 1998;**84**:983–991. DOI: 10.2307/3284632.

[25] Dubey JP, Lindsay DS, Saville WJA, Reed SM, Granstrom DE, Speer CA. A review of Sarcocystis neurona and equine protozoal myeloencephalitis (EPM). Veterinary Parasitology. 2001;**95**:89–131. DOI: 10.1016/S0304-4017(00)00384-8.

[26] Lobato J, Silva DA, Mineo TW, Amaral JD, Segundo GRS, Costa-Cruz JM, Mineo JR. Detection of immunoglobulin G antibodies to Neospora caninum in humans: High seropositivity rates in patients who are infected by human immunodeficiency virus or have neurological disorders. Clinical and Vaccine Immunology. 2006;**13**(1):84–89. DOI: 10.1128/CVI.13.1.84-89.2006.

[27] Mcallister MM, Dubey JP, Lindsay DS, Jolley WR, Wills RA, Mcguire AM. Dogs are definitive hosts of *Neospora caninum*. International Journal for Parasitology. 1998;**28**:1473–1478. DOI: 10.1016/S0020-7519(98)00138-6.

[28] Gondim LFP, Mcallister MM, Pitt WC, Zemlicka DE. Coyotes (Canis latrans) are definitive hosts of *Neospora caninum*. International Journal for Parasitology. 2004;**34**:159–161. DOI: 10.1016/J.IJPARA.2004.01.001.

[29] King JS, Slapeta J, Jenkins DJ, Al-Qassab SE, Ellis JT, Windsor PA. Australian dingos are definitive hosts of *Neospora caninum*. International Journal for Parasitology. 2010;**40**:945–950. DOI: 10.1016/J.IJPARA.2010.01.008.

[30] Dubey JP, Jenkins MC, Ferreira LR, Choudharya S, Verma SK, Kwok OCH, et al. Isolation of viable *Neospora caninum* from brains of wild gray wolves (Canis lupus). Veterinary Parasitology. 2014;**201**:150–153. DOI: 10.1016/J.VETPAR.2013.12.032.

[31] Poli A, Mancianti F, Carli MA, Stroscio MA, Kramer L. *Neospora caninum* infection in a Bernese catle dog from Italy. Veterinary Parasitology. 1998;**78(2)**:79–85. DOI: 10.1016/S0304-4017(98)00135-6.

[32] Figueredo LA, Dantas-Torres F, de Faria EB, Gondim LFP, Simões-Mattos L, Brandão-Filho SP, Mota RA. Occurrence of antibodies to *Neospora caninum* and *Toxoplasma gondii* in dogs from Pernambuco, Northeast Brazil. Veterinary Parasitology. 2008;**157(1)**:9–13. DOI: 10.1016/J.VETPAR.2008.07.009.

[33] Cañon-Franco WA, Bergamaschi DP, Labruna MB, Camargo LMA, Souza SLP, Silva JCR, et al. Prevalence of antibodies anti-*Neospora caninum* in dogs from Amazon, Brazil. Veterinary Parasitology. 2003;**115**:71–74. DOI: 10.1016/S0304-4017(03)00131-6.

[34] Hemphill A, Gottstein B, Kaufmann H. Adhesion and inavasion of bovine endothelial cells by *Neospora caninum*. Parasitology. 1996;**112**:183–197. DOI: 10.1017/S0031182000084754.

[35] Gonçalez CC, Paes AC, Langoni H, da Silva RC, Greca H, Camossi LG, Guimarães FF, Ullmann LS. Antibodies for Leptospira spp., Toxoplasma gondii and Neospora caninum in stray dogs housed in private kennel. Arquivo Brasileiro de Medicina Veterinaria e Zootecnia. 2010;**62(4)**:1011–1014. ISSN: 1011–1014.

[36] Bartels CJ, Wouda W, SchukkenYH. Risk factors for Neospora caninum associateal abortion astorms in dary herds in the Netherlands (1997 to 1997). Theriogenology. 1999;**52**:247–257. DOI: 10.1016/S0093-691X(99)00126-0.

[37] Farias NAR. Neosporosis: A disease being studied. Ciência e Tecnologia Veterinária. 2002;**1(1)**:5–14.

[38] Mcinnes LM, Irwin P, Palmer DG, Ryan UM. In vitro isolation and characterization of the first canine *Neospora caninum* isolate in Australia. Veterinary Parasitology. 2006;**137**:355–363. DOI: 10.1016/J.VETPAR.2006.01.018.

[39] Romanelli PR, Freire RL, Vidotto O, Marana ERM, Ogawa L, De Paula VSO, Navarro IT. Prevalence of *Neospora caninum* and *Toxoplasma gondii* in sheep and dogs from Guarapuava farms, Paraná State, Brazil. Research in Veterinary Science. 2007;**82(2)**:202–207. DOI: 10.1016/J.RVSC.2006.04.001.

[40] Moskwa B, Pastusiak K, Bien J, Cabaj W. The first detection of *Neospora caninum* DNA in the colostrum of infected cows. Parasitology Research. 2007;**100(3)**:633–636. DOI: 10.1007/S00436-006-0288-7.

[41] Dubey JP. Recents advances in *Neospora* and neosporosis. Veterinary Parasitology. 1999;**84 (3–4)**:349–367. DOI: 10.1016/S0304-4017(99)00044-8.

[42] Melo CB, Leite RC, Leite RC. Infection with Neospora caninum in dogs and other carnivores. Revista Conselho Federal de Medicina Veterinária, Brasília. 2005;**11(35)**:32–43.

[43] Gondim LFP, Gao L, McAllister MM. Improved production of Neospora caninum oocysts, cyclical oral transmission between dogs and cattle, and in vitro isolation from oocysts. Journal of Parasitology. 2002;**88(6)**:1159–1163. DOI: 10.1645/0022-3395(2002)088[1159:IPONCO]2.0.CO;2.

[44] Hemphill A, Gottstein B, Conraths FJ, de Meerschman F, Ellis JT, Innes EA, McAllister MM, Ortega-Moura LM, Tenter AJ, Trees AJ, Uggla A, Willians DJL, Wouda WA. European perspective on *Neospora caninum.* International Journal for Parasitology. 2000;**30**:877–924. DOI: 10.1016/S0020-7519(00)00072-2.

[45] Barber JS, Trees AJ. Clinical aspects of 27 cases of neosporosis in dogs. Veterinary Record. 1996;**139**:439–443. DOI: 10.1136/VR.139.18.439.

[46] Hill DE, Liddel S, Jenkins MC, Dubey JP. Specific detection of *Neospora caninum* oocysts in fecal samples from experimentally-infected dogs using the polymerase chain reaction. The Journal of Parasitology. 2001;**87**:395–398. DOI**:** 10.1645/0022-3395(2001)087[0395:SDONCO]2.0.CO;2.

[47] Slapeta JR, Modrý D, Kyselová I, Horejs R, Lukes J, Koudela B. Dogs shedding oocysts of *Neospora caninum*: PCR diagnosis and molecular phylogenic approach. Veterinary Parasitology. 2002;**109**:157–167. DOI: 10.1016/S0304-4017(02)00273-X.

[48] Atkinson R, et al. Progress in the serodiagnosis of *Neospora caninum* infection of cattle. Parasitology Today, Amsterdam. 2000;**16**:110–114. DOI: 10.1016/S0169-4758(99)01604-X.

[49] Bjorkman C, Uggla A. Serological diagnosis of *Neospora caninum* infection. International Journal for Parasitology, Oxford. 1999;**29(10)**:497–507. DOI: 10.1016/S0020-7519(99)00115-0.

[50] Barber JS, Trees AJ. Naturally occunning vertical transmission of Neospora caninum in dogs. Internacional Journal Parasitology. 1998;**28**:57–64. DOI: 10.1016/S0020-7519(97)00171-9.

[51] Maley SW, Buxton D, Macaldowie CN, Anderson IE, Wright SE, Bartley PM, Innes EA. Characterization of the immune response in the placenta of cattle experimentally infected with Neospora caninum in early gestation. Journal of Comparative Pathology. 2006;**135 (2)**:130–141. DOI: 10.1016/J.JCPA.2006.07.001.

[52] Thate FM, Laanen SC. Successful treatment of neosporosis in an adult dog. Veterinary Quarterly, Holanda. 1998;**20(Suppl 1)**:113–114. DOI: 10.1080/01652176.1998.10807458.

[53] Dijkstra T, Eysker M, Schares G, Conraths FJ, Wouda W, Barkema HW. Dogs shed Neosporacaninum oocysts after ingestion of naturally infected bovine placenta but not after ingestion of colostrum spiked with Neosporacaninum tachyzoites. International Journal for Parasitology. 2001;**31(8)**:747–752. DOI: 10.1016/S0020-7519(01)00230-2.

[54] Evermann JF, Ledbetter EC, Maes RK. Canine reproductive, respiratory, and ocular diseases due to canine herpesvirus. Veterinary Clinics of North America: Small Animal Practice. 2011;**41**:1097–1120. DOI: 10.1016/j.cvsm.2011.08.007.

[55] Megid J, Souza TD. Herpes vírus Canine herpesvirus. In: Megid J, Ribeiro MG, Paes AC, editors. infectious diseases in companion animals. de companhia. 1st ed. São Paulo: Roca; 2016. pp. 700–707. ISBN: 978-85-277-2789-1.

[56] Carmichael LE, Squire RA, Krook L. Clinical and pathologic features of a fatal viral disease of newborn pups. American Journal of Veterinary Research. 1965;**26**:803–814.

[57] ICTV (Org.). International Committee on Taxonomy of Viruses [Internet]. 2016. Available from: http://www.ictvonline.org/virustaxonomy.asp [Accessed: 2016-07-19].

[58] Greene CE. Infecção pelo Herpes-vírus Canine herpesvirus infection. In: EC Greene, editor. Infectious diseases in dogs and cats. 4th ed. São Paulo: Roca; 2015. pp. 50–56. ISBN 978-85-277-2690.

[59] Buonavoglia C, Martella V. Canine respiratory viruses. Veterinary Research. 2007;**30**:355–373. DOI: 10.1051/vetres:2006058.

[60] Dahlbom M, Johnsson M, Myllys V, Taponen J, Andersson M. Seroprevalence of canine herpesvirus-1 and Brucella canis in Finnish breeding kennels with and without reproductive problems. Reproduction in Domestic Animals. 2009;**44**:128–131. DOI: 10.1111/j.1439-0531.2007.01008.x.

[61] Rijsewijk FAM, Luiten EJ, Daus FJ, van der Heijden RW, van Oirschot JT. Prevalence of antibodies against canine herpesvirus 1 in dogs in the Netherlands in 1997–1998. Veterinary Microbiology. 1999;**65**:1–7. DOI: 10.1016/S0378-1135(98)00285-5.

[62] Ronsse V, Verstegen J, Thiry E, Onclin K, Aeberlé C, Brunet S, Poulet H. Canine herpesvirus-1 (CHV-1): Clinical, serological and virological patterns in breeding colonies. Theriogenology. 2005;**64**:61–74. DOI: 10.1016/j.theriogenology.2004.11.016.

[63] Monteiro FL, Cargnelutti JF, Martins M, Anziliero D, Erhardt MM, Weiblen R, Flores EF. Detection of respiratory viruses in shelter dogs maintained under varying environmental conditions. Brazilian Journal of Microbiology. 2016. In press. DOI: 10.1016/j.bjm.2016.07.002.

[64] Fernandes SC, Coutinho SDA. Canine infectious tracheobronchitis – review. Revista do Instituto de Ciências da Saúde. 2004;**22**:279–285.

[65] Carmichael L. Neonatal viral infections of pups: Canine herpesvirus and minute virus of canines (Canine Parvovirus-1). In: Carmichael L, editor. Recent Advances in Canine Infectious Diseases. Ithaca: International Veterinary Information Service; 2004. Available from: www.ivis.org/.

[66] Ledbetter EC. Canine herpesvirus-1 ocular diseases of mature dogs. New Zealand Veterinary Journal. 2013;**61**:193–201. DOI: 10.1080/00480169.2013.768151.

[67] Lamm CG, Njaa BL. Clinical approach to abortion, stillbirth, and neonatal death in dogs and cats. Veterinary Clinics of North America. Small Animal Practice. 2012;**42**:501–513. DOI: 10.1016/j.cvsm.2012.01.015.

[68] Carmichael LE, Schlafer DH, Hashimoto A. Minute virus of canine (MCV, canine parvovirus type-1): Pathogenicity for pups and seroprevalence estimate. Journal of Veterinary Diagnostic Investigation. 1994;**6**:165–174. DOI: 10.1177/104063879400600206.

[69] Paes AC. Parvovirose canina. In: Megid J, Ribeiro MG, Paes AC, editors. Doenças infecciosa em animais de companhia. 1st ed. São Paulo: Roca; 2016.

[70] Decaro N, Carmichael LE, Buonavoglia C. Viral reproductive pathogens of dogs and cats. Veterinary Clinics of North America: Small Animal Practice. 2012;**42**:583–598. DOI: 10.1016/j.cvsm.2012.01.006.

[71] Manteufel J, Truyen U. Animal bocaviruses: A brief review. Intervirology. 2008;**51**:328–334. DOI: 10.1159/000173734.

[72] Decaro N, Altamura M, Pratelli A, Pepe M, Tinelli A, Casale D, Martella V, Tafaro A, Camero M, Elia G, Tempesta M, Jirillo E, Buonavoglia C. Evaluation of the innate immune response in pups during canine parvovirus type 1 infection. New Microbioligica. 2002;**25**:291–298.

[73] Mochizuki M, Hashimoto M, Hajima T, Takiguchi M, Hashimoto A, Une Y, Roerink F, Ohshima T, Parrish CR, Carmichael LE. Virologic and serologic identification of minute virus of Canine viral enteritis. In: EC Greene, editor. Infectious diseases of dogs and cats. in Japan. Jounal of Clinical Microbiology. 2002;**40**:3993–3998. DOI: 10.1128/JCM.40.11.3993-3998.2002.

[74] Greene CE, Decaro N. Enterite viral canina In: Greene CE, editor. Doenças infecciosas em cães e gatos. 4th ed. São Paulo: Roca; 2015. pp. 69–79. ISBN: 978-85-277-2690.

[75] Carmichael LE, Schlafer DH, Hashimoto A. Pathogenicity of minute virus of canines (MVC) for the canine fetus. Cornell Veterinarian. 1991;**81**:151–171.

[76] Oura CA, El Harrak M. Midge-transmitted bluetongue in domestic dogs. Epidemiology & Infection. 2011;**139**:1396–1400. DOI: 10.1017/S0950268810002396.

[77] Dubovi EJ, Hawkins M, Griffin RA, Johnson DJ, Ostlund EN. Isolation of Bluetongue virus from canine abortions. Journal of Veterinary Diagnostic Investigation. 2013;**25**:490–492. DOI: 10.1177/1040638713489982.

[78] Gaudreault NN, Jasperson DC, Dubovi EJ, Johnson DJ, Ostlund EN, Wilson WC. Whole genome sequence analysis of circulating Bluetongue virus serotype 11 strains from the United States including two domestic canine isolates. Journal of Veterinary Diagnostic Investigation. 2015;**27**:442–448. DOI: 10.1177/1040638715585156.

[79] Krakowka S, Hoover EA, Koestner A, Ketring K. Experimental and naturally occurring transplacental transmission of canine distemper virus. American Journal of Veterinary Research. 1977;**38**:919–922.

[80] Verstegen J, Dhaliwal G, Verstegen-Onclin K. Canine and feline pregnancy loss due to viral and non-infectious causes: A review. Theriogenology. 2008;**70**:304–319. DOI: 10.1016/j.theriogenology.2008.05.035.

[81] Johnston SD, Root Kustritz MV, Olson PNS. Canine pregnancy. In: Kersey R, editor. Canine and Feline Theriogenology. 1st ed. Philadelphia: W.B. Saunders Company; 2001. pp. 66–104. ISBN: 978–0721656076.

[82] Greene CE, Appel MJ. Canine distemper. In: Greene CE, editor. Infectious Diseases of the Dog and Cat. 3rd ed. St. Louis: Elsevier; 2006. pp. 25–41.

[83] Almes KM, Janardhan KS, Anderson J, Hesse RA, Patton KM. Fatal canine adenoviral pneumonia in two litters of bulldogs. Journal of Veterinary Diagnostic Investigation. 2010;**22**:780–784. DOI: 10.1177/104063871002200524.

[84] Rossi GL, Pauli B, Luginbühl H, Probst D. Demonstration by electron microscopy of viruses in cells found by light microscopy to contain inclusion bodies. Indian Journal of Pathology and Microbiology (Basel). 1972;**38**:321–332. DOI: 10.1159/000162432.

Section 2

Oncology

Chapter 4

Mast Cell Tumors

Yosuke Amagai and Akane Tanaka

Additional information is available at the end of the chapter

http://dx.doi.org/10.5772/65797

Abstract

Mast cell tumor is one of the major cutaneous tumors in dogs. Though the etiology of MCTs is not completely understood, it becomes clear that approximately 10–20% MCTs express mutant KIT receptors with ligand-independent phosphorylation. Tyrosine kinase inhibitors targeting KIT exert antitumor effects on malignant proliferation of mast cells with or without gene mutations. However, the efficacy of KIT inhibitors on dogs with MCTs has been limited. In this chapter, we would like to outline the general understandings of mast cells such as the process of its differentiation and proliferation, and what has been revealed regarding the mechanism of tumorigenesis and therapeutic approaches. In particular, KIT mutation-related evidences and therapeutic approaches in the future are discussed.

Keywords: mast cell tumor, stem cell factor, KIT, tyrosine kinase inhibitor

1. Introduction

Mast cells are originated from hematopoietic stem cells in bone marrow as similar to other granulocytes. However, mast cells are very unique because they are released from bone marrow as undifferentiated progenitor cells to the circulation, and their final maturation is completed in the peripheral tissues. We find mast cells in most of the tissues and organs in our body, particularly in interstitial connective tissues of each organ being close to blood vessels. Mast cells play crucial roles in innate immunity against parasites and microbes that is essential for host defense in humans and animals. In acquired immunity, activation of mast cells is induced by cross-linkage of IgE that binds to high-affinity IgE receptors after antigen exposure. Moreover, some chemicals and toxins as well as physical (i.e., scratching and heat) and neurogenic stimuli trigger activation of mast cells. Various chemical mediators and cytokines are released from mast cells after their degranulation, and sometimes initiate allergic inflammation and itch sensation. In canine medicine, serious involvement of mast cells in allergic

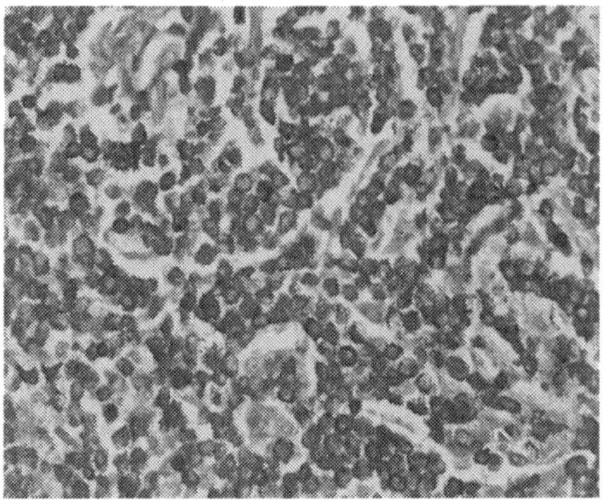

Figure 1. Typical histologic features of surgically removed canine MCTs stained with toluidine blue solution. A mass is consisted with numbers of differenced mast cells that show metachromasia. Connective tissues colored in light blue can be identified.

diseases has been identified, and mast cells are estimated as the most important target in medical treatment of allergy. Although malignancies of mast cell are uncommon disorders in humans, veterinary clinicians frequently encounter mast cell tumors (MCTs) in dogs and cats. The frequency of cutaneous MCTs has been reported to reach 20% of all tumors raised in the skin of dogs. Most of the malignant mast cells existed in a mass have granules in their cytosol, containing pruritogens, inflammatory factors, and various proteases (**Figure 1**). In this chapter, recent information on both basics in mast cell biology and clinical approaches for canine MCTs is outlined.

2. Differentiation and proliferation of mast cells

2.1. Differentiation of mast cells

Mast cell progenitors are differentiated from pluripotent hematopoietic stem cells in bone marrow. Being different from other granulocytes, mast cells are transported by blood stream to peripheral tissues as mononuclear immature cells without granules in their cytosol [1]. In peripheral tissues, mast cell precursors complete their differentiation and distribution being ready for the host defense (**Figure 2**). According to characters of microenvironment of each peripheral tissue, final types of mature mast cells are altered. In the skin, they differenced into connective tissue-type mast cells that include heparin proteoglycan and abundant granules in cytosol. Various kinds of proteases and chemical mediators are in granules of connective tissue-type mast cells by which sever inflammation is induced when they are released at the affected sites. Heparin proteoglycan-positive mast cells can be detected with not only toluidine blue but also berberine sulfate and safranin O. In contrast, mast cells that reach to and invade in mucosal tissues differentiate into mucosal-type mast cells. Mucosal-type mast cells include chondroitin sulfate as proteoglycan, and cytosolic granules are very few. Chondroitin

http://dx.doi.org/10.5772/65797

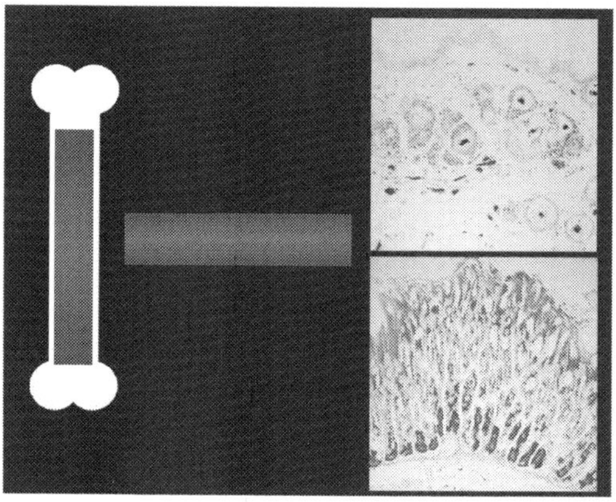

Figure 2. Origin and transportation of mast cell precursors. Immature mast cells without cytosolic granules are released from bone marrow and distributed to peripheral tissues via circulation. Final maturation of mast cells depends on factors and molecules expressed in microenvironment of each tissue. In the skin (the upper panel), connective tissue-type mast cells with rich cytosolic granules are differentiated. In mucosa of the gastrointestinal tract (the lower panel), mucosal-type mast cells with few cytosolic granules are differentiated.

sulfate-positive mast cells are identified with alcian blue staining, but not with toluidine blue, berberine sulfate, or safranin O. Connective tissue-type mast cells are found in the skin and connective tissues in various organs. On the other hand, mucosal-type mast cells are differentiated in mucosa of the gastrointestinal tract.

2.2. Proliferation of mast cells

Factors that involve in mast cell proliferation are not fully understood. In mice, mast cells proliferate in response to interleukin (IL)-3, IL-9, and stem cell factor (SCF). Both IL-6 and SCF are essential factors for proliferation and survival of human mast cells. Though less is known regarding development of canine mast cells, SCF has been reported to induce proliferation and survival, as well as migration and activation of canine mast cells [2, 3]. The IL dependency varies according to a species from which mast cells are derived. However, SCF is the most important factor for proliferation, differentiation, and survival of mast cells in humans and animals. SCF is the ligand for KIT receptors expressed on cell surface of mast cells. Since mast cells are differentiated from KIT-positive cell lineage and KIT is retained on the surface of not only immature precursors but also mature mast cells, influence of SCF on mast cells must be crucial for their proliferation and differentiation. In fact, gain-of-function mutations in the *KIT* gene have been identified in mast cells with factor-independent tumorigenic proliferation in humans, rodents, and dogs. Meanwhile, over 50% of canine MCTs and most of the feline MCTs show *KIT*gen mutation-independent development. Recent observations have revealed that not only SCF but also other cytokines and growth factors including epidermal growth factor, vascular endothelial growth factor, and nerve growth factor may facilitate proliferation, differentiation, and survival of mast cells. Unfortunately, investigation on development of canine mast cells has been insufficient.

3. Cellular biology of mast cells

3.1. Inflammatory responses of mast cells

Mast cells play key roles for inflammatory responses through degranulation and cytokine/chemokine production and secretion [4]. However, proteases, chemical mediators, and cytokines included in mast cell granules are different according to types of mast cells. Heterogeneity among mast cells is important to precisely understand the pathophysiological roles of mast cells in each tissue. Connective tissue-type mast cells include chemical mediators such as histamine that has strong biologic activity on nerve fibers and blood vessels. Histamine is known as a pruritogen that induces itch sensation resulting in scratching behavior in humans and animals. Vasoactive effects of histamine initiate inflammation at the affected sites. Scratching behavior stimulates physical degranulation of mast cells leading to exacerbation of swelling and inflammation. Exact roles of mucosal-type mast cells are not clearly demonstrated. Since numbers of mucosal-type mast cells are increased in the gut with parasite infection, roles in host reaction against parasite exclusion have been proposed. However, factors and mechanisms that involve in antiparasite effects of mucosal-type mast cells are not fully explored.

3.2. Heterogeneity of MCTs

Most MCTs in dogs are consisted with histamine-rich connective tissue-type mast cells. Steps responsible for transformation of mast cells are summarized in **Figure 3**. Since receptors for histamine are expressed in mucosal cells of the stomach, progression of MCTs has been suggested to induce gastric ulcer and serious damage on gastric function. However, the exact and direct association of MCTs with gastric ulcer remains unclear [5, 6]. Connective tissue-type mast cells contain tryptase and chymase that possess broad protease activities. Particularly, tryptase has been reported to regulate neovascularization. Connective tissue-type mast cells can also produce growth factors for vascular endothelial cells; therefore,

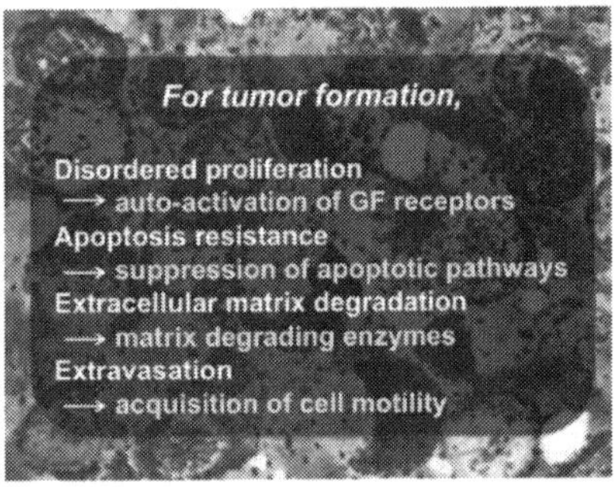

Figure 3. Malignant transformation of mast cells. For tumor formation, mast cell proliferation must be promoted by activation of ligand-independent activation on growth factor receptors. Also, acquisition of resistance against apoptosis pathways and invasive characters may facilitate malignant expansion of MCTs.

most of the MCTs must be rich with blood vessels. Moreover, dogs with serious MCTs show deficiency in blood coagulation possibly because connective tissue-type mast cells have plenty of heparin in their cytosol. Canine mast cells have unique chymase whose name is dog mast cell protease-3; however, its specific role has not been fully understood [7]. A mass formed with mucosal-type mast cells rarely develops in the gastrointestinal tract, which induces dysfunction of the gut. Influence of mucosal mast cell tumor on other organs except the gut has not been well documented.

4. Pathology and diagnosis of MCTs

4.1. Clinical presentation, incidence, and risk factors

MCTs are characterized by the aberrant proliferation of mast cells, accounting for approximately 20% of cutaneous tumors in dogs [8, 9]. They develop in the subcutaneous tissue and dermis in most cases, and other types of mast cell malignancies such as mast cell leukemia and visceral MCTs are rarely observed. They usually occur in the trunk or limbs and sometimes observed in the head and neck (**Figure 4**) [9, 10]. Because mast cells release proinflammatory mediators, erythema and wheal called Darier's sign are sometimes observed. However, MCT-specific symptoms that can distinguish MCTs from other tumors are rare.

Risk factors for MCTs, age, sex, breed, spay/castration, and tumor grading have been reported [8, 11]. Among these, most factors except sex are deeply related to its incidence [11]. Recently, genetic characteristics have been also investigated, showing the high correlation of KIT mutations with prognosis of dogs with MCTs [12–14]. As Mochizuki et al. [11] presented nice summary on each factor except genetic one, we would like to focus on the genetic characteristics of MCTs in the following sections.

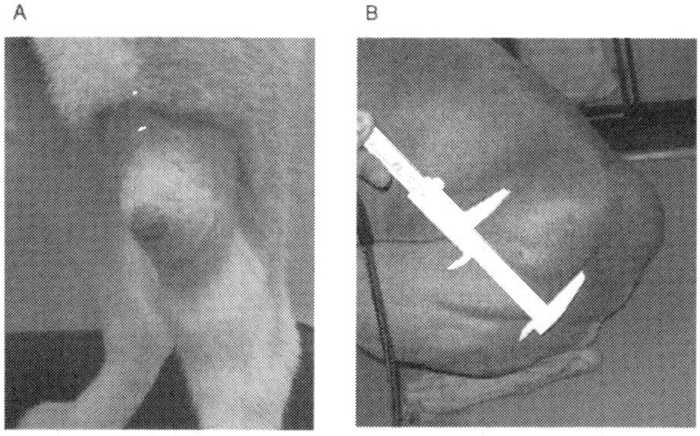

Figure 4. Representative photograph of MCT-diagnosed dog. (A) MCT occurred in the left leg in 11-year-old female, Shiba. (B) MCT occurred in the left waist in 8-year-old female, Pharaoh Hound.

4.2. Diagnosis of MCTs

Cytological or histological analyses through a fine needle aspiration or biopsy are required for the diagnosis of MCTs. Typically, round-shape cells with round nuclei and with rich cytosol are observed. Mast cells have abundant cytosolic granules, and specific staining methods with toluidine blue or safranin O can identify them. However, MCTs with undifferentiated more malignant mast cells possess few granules. Confirmation of the swelling of draining lymph nodes and sometimes fine needle aspiration of the lymph node may help to determine the presence of metastasis. Patnaik grading is mainly used pathological grading in the veterinary field because it is recognized as a good prognostic marker [15]. There are three pathological grades (grades I, II, and III), and higher grade indicates that the tumor is more malignant [15]. Several analyses have been revealed: the correlation between tumor grading and the c-*kit* gene mutation, clearly showing that c-*kit* mutations are more frequently observed in high-grade tumor [12–14]. Therefore, analyzing c-*kit* sequence can also be a prognostic marker for MCTs. In addition, analyzing c-*kit* gene is important in terms of selecting proper treatments because several molecular target inhibitors against KIT protein, a receptor encoded in the c-*kit* gene, are currently available for the treatment of MCTs. Polymerase chain reaction that amplifies exon 11 and intron 11 region using tumor genome enables the detection of internal tandem duplications (ITDs) in the juxtamembrane domain, which is the most frequent type of c-*kit* mutation. Recently, however, whole sequence of c-*kit* mRNA is more common because the proportion of mutations in other region of KIT domain or other type of mutations in the juxtamembrane domain are not negligible. Recent reduction in the cost for sequencing analysis will probably boost this trend (see Section 5).

5. Neoplastic transformation of mast cells

5.1. Overview

Mast cell malignancies are observed among species, though the incidence of mast cell malignancies is much lower in human and rodents than in dogs [8, 9]. One of the well-investigated mechanisms of mast cell tumorigenesis is mutations in the c-*kit* gene [9, 10]. We would like to overview the current understandings of the mutant KIT contribution on mast cell tumorigenesis as well as other tumor-related transformations of mast cells that may correlate with their tumorigenesis in mast cells.

5.2. KIT mutation-dependent neoplastic transformation

5.2.1. KIT mutations in human and rodents

KIT is a type III receptor tyrosine kinase of which ligand is stem cell factor (SCF) (**Figure 5**). It is consisted of the extracellular domain, transmembrane domain, juxtamembrane domain, and tyrosine kinase domain [16] (**Figure 5**). In human, aberrant proliferation of mast cell is observed in the patients of systemic/cutaneous mastocytosis, mast cell sarcoma, and mast cell leukemia [17]. Among them, most systemic mastocytosis occurs due to the mutations in the tyrosine kinase domain of KIT, especially a point mutation in Asp816 [12]. In general, SCF binding to KIT triggers the conformational changes in KIT and leads to the dimerization of

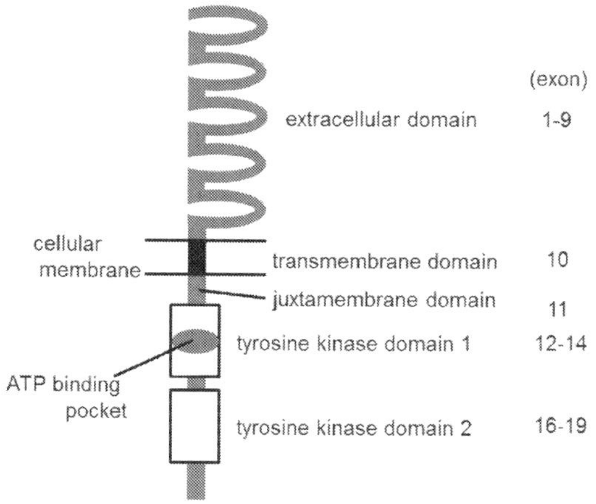

Figure 5. Schematic diagram of KIT protein. The numbers of exon correspond to the ones in dog KIT.

the protein, allowing the binding of adenosine triphosphate (ATP) and phosphorylation of tyrosine kinase domain [18]. Mutations in the tyrosine kinase domain alter the conformation to the one similar to its active form, thus resulting in the constitutive KIT activation even in the absence of either SCF binding, KIT dimerization, or ATP binding [19]. Neoplastic growth of mast cells is rarely observed in rodents, through currently available rodent-derived mast cell lines. For example, RBL-2H3 cells (derived from a Wistar rat) and P815 cells (derived from a DBA/2 mouse) express Asp817Tyr and Asp814Tyr, respectively, which correspond to the Asp816 mutation in human KIT [20]. As far as we know, other mechanisms that trigger rodent mast cell tumorigenesis have not been reported.

5.2.2. KIT mutations in dog

In contrast to human and rodents, KIT mutations in dog MCT are frequently observed in the juxtamembrane domain (**Table 1**). ITDs in the domain are the first discovered and the most frequent mutations in canine MCTs [21] (**Table 1**). Besides ITDs, other mutations in the juxtamembrane domain or extracellular domain have been also reported [21]. We recently demonstrated that most of these mutations in the extracellular or juxtamembrane domain cause aberrant KIT activation and neoplastic proliferation of mast cells by triggering ligand-independent dimerization (Ref. [22] and unpublished data). In contrast to the mutations in the tyrosine kinase domain, these mutations require ATP for the phosphorylation of the tyrosine kinase domain, providing a rationale for using ATP-competitive small molecule inhibitors for suppressing the aberrant KIT activations [18, 23].

5.3. KIT mutation-independent neoplastic transformation

Few mechanisms of mast cell tumorigenesis except KIT mutations have been identified, but we recently demonstrated that MCT cells produce SCF and support their growth in a

Analyzed exons	Domain	Exon	Mutation	Frequency (%)	References
All	Extracellular	8	417-421ITD	8/202 (4.2%)	[24]
		8	Gln430Arg	1/202 (0.5%)	
		9	Ser479Ile	5/202 (2.5%)	
		9	Asn508Ile	3/202 (1.5%)	
	Juxtamembrane	11	555–557mutation	7/202 (3.5%)	
		11	ITD	25/202 (12.4%)	
	Tyrosine kinase	17	del826–828	1/202 (0.5%)	
All	Extracellular	6, 7, 8	del102AA, Thr414Ala	1/47 (2.1%)	[25]
		8	417-420ITD	3/47 (6.4%)	
	Extracellular, transmembrane	9,10	ins4AA	1/47 (2.1%)	
	Juxtamembrane	11	del558	1/47 (2.1%)	
		11	Leu575Pro	1/47 (2.1%)	
		11	ITD	8/47 (17.0%)	
	Tyrosine kinase	15	del714	1/47 (2.1%)	
11	Juxtamembrane	11	ITD	8/88 (9.1%)	[12]
		11	del	4/88 (4.5%)	
11	Juxtamembrane	11	ITD	8/60 (13.3%)	[13]
11	Juxtamembrane	11	ITD	7/68 (10.3%)	[14]
11	Juxtamembrane	11	ITD	24/118 (20.3%)	[26]
8, 9, 11	Extracellular	8	421-429del, ins3AA	1/21 (4.8%)	[27]

AA, amino acid.

Table 1. KIT mutations that have been reported in dog MCTs.

paracrine/autocrine manner [28]. In the analyses, high SCF production was confirmed in multiple clinical MCT samples [29]. It may explain the high response of clinical MCTs to KIT-specific molecular inhibitors even when the tumor cells express wild-type KIT. This will be further discussed in the following section.

Recent approaches such as next-generation sequencing will reveal even minor mutations or single nucleotide polymorphisms in neoplastic mast cells. In fact, Spector et al. [30] and Youk et al. [31] discovered a human mast cell leukemia-specific mutation in several genes. As the cost of these approaches decreases, they will be introduced in the veterinary field, probably leading to the deep understanding of mast cell tumorigenesis among species. Another approach aiming at the control of tumor growth is modifying epigenetic status in tumor genome [32]. Regarding an epigenetic alteration in MCTs, Morimoto et al. [33] showed that DNA hypomethylation widely occurred in malignant, higher-grade MCTs. Moreover, antitumor effects of AR-42, a histone deacetylase inhibitor, on several MCT cell

http://dx.doi.org/10.5772/65797

lines as well as primary tumor cells have been demonstrated [34]. Thus, the characterization of epigenetic alteration is likely to be an effective approach to reveal MCT transformation.

6. Treatment and prognosis

6.1. Conventional therapies

Complete surgical removal with wide margin is the best way for solitary cutaneous MCTs. However, for dogs with multiple masses, metastasis, or sever invasive MCTs, surgical removal is sometimes incompetent. Chemotherapies have failed to overcome MCTs. However, to reduce tumor size, some combination chemotherapies have applied. Glucocorticoid is one of the most important drugs for treatment of MCTs. More than 70% of cutaneous MCTs in dogs respond well to oral administration of glucocorticoid [35–37]. Expression levels of glucocorticoid receptors in MCT cells have been reported to associate with glucocorticoid sensitivity [37]. Glucocorticoid shows strong antitumor effects on MCT cells with high expression of glucocorticoid receptors. Oral administration of glucocorticoid must be an easy and effective chemotherapy for canine MCTs. Since side effects induced by glucocorticoid administration will sometimes be concerned, clinicians must pay attention on blood chemistry data and general conditions of dog patients. Glucocorticoid is usually applied as a part of multidrug chemotherapies for MCTs. Anticancer drugs, such as vincristine, vinblastine, cyclophosphamide, and CCNU, have been tested for combination chemotherapies for MCTs with or without glucocorticoid [38–42]. However, very little information on the chemotherapeutic response of MCTs can be obtained. Since recent studies have been based on small numbers of cases and have often included MCTs of different pathologic grades and clinical stages, data must be carefully evaluated [43]. Recently, adjuvant chemotherapies have been proposed in treatment of various cancers and sarcomas. Since neo-adjuvant administration with glucocorticoid usually reduces mass size of MCTs, wide surgical margins will be obtained [37]. On the other hand, postoperative adjuvant chemotherapy is suggested to kill MCT cells that remain at the affected site after incomplete excision. Several trials on adjunctive chemotherapy have been reported. However, most of the adjunctive chemotherapy does not appear to increase survival times. Although surgical removal with radiation has been tested for MCTs, remarkable improvement is not provided. No difference in overall survival rate has been observed between dogs with MCTs receiving and not receiving prophylactic irradiation of the regional lymph node [44].

6.2. Molecular target therapies targeting KIT

Because aberrant activation of mutant KIT is one of the causes in mast cell tumorigenesis, anticancer effects of KIT inhibitors have been investigated. In fact, clinical trials that enrolled MCT-diagnosed dogs have been undertaken to evaluate the efficacy of molecular targeting agents against KIT. We would like to overview the history of the research on KIT inhibitors and discuss therapeutic perspective fwor MCTs.

	All MCTs		MCTs with mutant KIT		MCTs without mutant KIT	
	Cases	CR + PR	Cases	CR + PR	Cases	CR + PR
Toceranib	86	32 (37.2%)	20	12 (60.0%)	66	20 (30.3%)
Placebo	63	5 (7.9%)	9	1 (11.1%)	54	4 (7.4%)
Placebo>toceranib	58	24 (41.4%)	9	7 (77.8%)	49	17 (34.7%)

The number of the enrolled patients and the proportion of CR/PR cases are indicated.

Table 2. Summary of a clinical trial of toceranib phosphate [26].

The first molecular inhibitor applied to human was the imatinib mesylate, which repress activations of KIT, platelet-derived growth factor receptor (PDGFR), and Bcr-Abl [45]. It was first administered to the patient of gastrointestinal stromal tumor with a mutation in the juxtamembrane domain of KIT [46]. At around the same time, KIT mutations in canine MCT were first discovered [21], suggesting the possibility that KIT inhibitors can be applied to dog MCTs. Actually, there have been several reports that show the inhibitory effects of imatinib mesylate on MCTs, especially for the tumor cells with KIT mutations in either the extracellular domain or juxtamembrane domain [22, 47]. Based on these results from basic researches, some clinicians administered imatinib mesylate to MCT-diagnosed dogs and obtained partial response at least in some of them [47]. However, there was no rationale for the administration of imatinib mesylate to MCT-diagnosed dogs through the clinical trials. In contrast to that, both masitinib mesylate and toceranib phosphate are approved by either the Food and Drug Administration (FDA) in the United States or European Medicines Agency based on the results from the clinical trials enrolling MCT-diagnosed dogs [26, 48,49]. Basically, imatinib mesylate, masitinib mesylate, and toceranib phosphate are all ATP-competitive inhibitors, and they suppress the activation of mutant KITs that require ATP binding for their activation.

Results of clinical trials for toceranib phosphate, which is a random double-blind trials, are summarized in **Table 2** [26]. In this trial, more than 150 MCT patients were enrolled. After the six-week treatment, either complete response (CR) or partial response (PR) was obtained 32 in 86 (37.2%) patients in the treatment group, while the proportion of CR/PR was only 7.9% (5 in 63 patients) in the placebo group. In addition, a group treated with toceranib phosphate following placebo-escape, which administered toceranib phosphate after the placebo treatment, responded the agent, resulting in the CR/PR in 24 cases out of 58 (41.4%). At least in this trial, significant increase in severe adverse effects (grade III or IV) was not detected. In case of mastinib mesylate, a phase III trial was carried out in France, enrolling more than 130 MCT patients. One-year survival and two-year survival were 62.1 and 39.8%, respectively, in the treatment group, though the ones were 36.0 and 15.0%, respectively, in the placebo group [49].

Interestingly, both agents showed antitumor effects even on MCTs expressing wild-type KIT (**Tables 2** and **3**). Though the agents suppress the activation of other receptor tyrosine kinases

	All MCTs		MCTs with mutant KIT		MCTs without mutant KIT	
	1 year	2 year	1 year	2 year	1 year	2 year
Toceranib	59/95 (62.1%)	33/83 (39.8%)	17/27 (63.0%)	7/23 (30.4%)	38/62 (61.3%)	23/54 (42.6%)
Placebo	9/25 (36.0%)	3/20 (15.0%)	1/7 (14.3%)	0/6 (0%)	7/16 (43.8%)	2/12 (16.7%)

The number and proportion of the patients that survived more than 1 or 2 years are indicated.

Table 3. Summary of a clinical trial of masitinib phosphate [49].

such as platelet-derived growth factor receptor or vascular endothelial growth factor receptor 2 [50, 51], aberrant activations of principal targets except KIT were not observed in our study using more than 30 clinical MCT tissue samples (unpublished data). Thus, it is likely that the data in **Table 2** indicate that tumor growth in no more than 30% of MCT was dependent on KIT signaling even though they express wild-type KIT. We consider that these can be at least partly explained by SCF autoproduction from tumor cells as described above [28, 29]. Though further investigations are necessary, analyses to determine the KIT activation status will probably be a direct diagnostic agent to accurately predict the therapeutic efficacy of KIT targeting inhibitors.

7. Final remarks

As described, molecular biological approaches to MCTs have started, and new findings are accumulating. However, some clinical researches present very limited information obtained from few cases. Therefore, clinicians should carefully collect information and evaluate data before clinical application of anticancer drugs and molecular targeting drugs to dogs with MCTs. It is dangerous to trust all data reported in few research reports or on few clinical cases. Knowledge on basic biology of mast cells will help clinicians to understand the recent molecular approaches to MCTs.

Author details

Yosuke Amagai[1, 2] and Akane Tanaka[3*]

*Address all correspondence to: akane@cc.tuat.ac.jp

1 Research Fellow of the Japan Society for the Promotion of Science, Tokyo, Japan

2 Tokyo Metropolitan Institute of Medical Science, Tokyo, Japan

3 Tokyo University of Agriculture and Technology, Tokyo, Japan

References

[1] Kitamura Y. Heterogeneity of mast cells and phenotypic change between subpopulations. Annu Rev Immunol. 1989;**7**:59–76. doi:10.1146/annurev.iy.07.040189.000423

[2] Brazís P, Queralt M, de Mora F, Ferrer LI, Puigdemont A. Stem cell factor enhances IgE-mediated histamine and TNF-alpha release from dispersed canine cutaneous mast cells. Vet Immunol Immunopathol. 2000;**75**:97–108.

[3] Lin TY, Rush LJ, London CA. Generation and characterization of bone marrow-derived cultured canine mast cells. Vet Immunol Immunopathol. 2006;**113**:37–52.

[4] Galli SJ, Tsai M. Mast cells: versatile regulators of inflammation, tissue remodeling, host defense and homeostasis. J Dermatol Sci. 2008;**49**:7–19.

[5] Howard EB, Sawa TR, Nielsen SW, Kenyon AJ. Mastocytoma and gastroduodenal ulceration: gastric and duodenal ulcers in dogs with mastocytoma. Pathol Vet. 1969;**6**:146–58.

[6] Ammann RW, Vetter D, Deyhle P, Tschen H, Sulser H, Schmid M. Gastrointestinal involvement in systemic mastocytosis. Gut. 1976;**17**:107–112.

[7] Raymond WW, Tam EK, Blount JL, Caughey GH. Purification and characterization of dog mast cell protease-3, an oligomeric relative of tryptases. J Biol Chem. 1995;**270**:13164–13170.

[8] Withrow and MacEwen's small animal clinical oncology. 5th ed. Elsevier Inc.; 2013. p. 335–355. DOI: 10.1111/avj.12086

[9] Welle MM, Bley CR, Howard J, Rüfenacht S. Canine mast cell tumours: a review of the pathogenesis, clinical features, pathology and treatment. Vet Dermatol. 2008;**19**:321–339. doi:10.1111/j.1365-3164.2008.00694.x.

[10] Blackwood L, Murphy S, Buracco P, De Vos J, Fornel-Thibaud D, Hirschberger J, Kessler M, Pastor J, Ponce F, Savary-Bataille K. European consensus document on mast cell tumours in dogs and cats. Vet Comp Oncol. 2012;**10**:e1–e29. doi:10.1111/j.1476-5829.2012.00341.x.

[11] Mochizuki H, Motsinger-Reif A, Bettini C, Moroff S, Breen M. Association of breed and histopathological grade in canine mast cell tumours. Vet Comp Oncol. DOI: 10.1111/vco.12225.

[12] Zemke D, Yamini B, Yuzbasiyan-Gurkan V. Mutations in the juxtamembrane domain of c-KIT are associated with higher grade mast cell tumors in dogs. Vet Pathol. 2002;**39**:529–535. doi:10.1354/vp.39-5-529

[13] Webster JD, Yuzbasiyan-Gurkan V, Kaneene JB, Miller R, Resau JH, Kiupel M. The role of c-KIT in tumorigenesis: evaluation in canine cutaneous mast cell tumors. Neoplasia. 2006;**8**:104–111. doi:10.1593/neo.05622

[14] Amagai Y, Tanaka A, Matsuda A, Jung K, Oida K, Nishikawa S, Jang H, Matsuda H. Heterogeneity of internal tandem duplications in the c-*kit* of dogs with multiple mast cell tumours. J Small AnimPract. 2013;**54**:377–380. doi:10.1111/jsap.12069

[15] Patnaik AK, Ehler WJ, MacEwen EG. Canine cutaneous mast cell tumor: morphologic grading and survival time in 83 dogs. Vet Pathol. 1984;**21**:469–474. doi:10.1177/030098588402100503

[16] Broxmeyer HE, Maze R, Miyazawa K, Carow C, Hendrie PC, Cooper S, Hangoc G, Vadhan-Raj S, Lu L. The kit receptor and its ligand, steel factor, as regulators of hemopoiesis. Cancer Cells. 1991;3:480–487.

[17] Akin C, Valent P. Diagnostic criteria and classification of mastocytosis in 2014. Immunol Allergy Clin North Am. 2014;**34**:207–218. doi:10.1016/j.iac.2014.02.003

[18] Mol CD, Lim KB, Sridhar V, Zou H, Chien EY, Sang BC, Nowakowski J, Kassel DB, Cronin CN, McRee DE. Structure of a c-*kit* product complex reveals the basis for kinase transactivation. J Biol Chem. 2003;**278**:31461–31464. doi:10.1074/jbc.C300186200

[19] Vendôme J, Letard S, Martin F, Svinarchuk F, Dubreuil P, Auclair C, Le Bret M. Molecular modeling of wild-type and D816V c-Kit inhibition based on ATP-competitive binding of ellipticine derivatives to tyrosine kinases. J Med Chem. 2005;**48**:6194–6201. doi:10.1021/jm050231m.

[20] Beghini A, Cairoli R, Morra E, Larizza L. In vivo differentiation of mast cells from acute myeloid leukemia blasts carrying a novel activating ligand-independent c-*kit* mutation. Blood Cells Mol Dis. 1998;**24**:262–270. doi:10.1006/bcmd.1998.0191

[21] London CA, Galli SJ, Yuuki T, Hu Z, Helfand SC, Geissler EN. Spontaneous canine mast cell tumors express tandem duplications in the proto-oncogene c-*kit*. ExpHematol. 1999;**27**:689–697. doi:10.1016/S0301-472X(98)00075-7

[22] Amagai Y, Matsuda A, Jung K, Oida K, Jang H, Ishizaka S, Matsuda H, Tanaka A. A point mutation in the extracellular domain of KIT promotes tumorigenesis of mast cells via ligand-independent auto-dimerization. Sci Rep. 2015;**5**. doi:10.1038/srep09775

[23] Mol CD, Dougan DR, Schneider TR, Skene RJ, Kraus ML, Scheibe DN, Snell GP, Zou H, Sang BC, Wilson KP. Structural basis for the autoinhibition and STI-571 inhibition of c-Kit tyrosine kinase. J Biol Chem. 2004;**279**:31655–31663. doi:10.1074/jbc.M403319200

[24] Letard S, Yang Y, Hanssens K, Palmerini F, Leventhal PS, Guery S, Moussy A, Kinet JP, Hermine O, Dubreuil P. Gain-of-function mutations in the extracellular domain of KIT are common in canine mast cell tumors. Mol Cancer Res. 2008;**6**:1137–1145. doi:10.1158/1541-7786.MCR-08-0067

[25] Takeuchi Y, Fujino Y, Watanabe M, Takahashi M, Nakagawa T, Takeuchi A, Bonkobara M, Kobayashi T, Ohno K, Uchida K. Validation of the prognostic value of histopathological grading or c-*kit* mutation in canine cutaneous mast cell tumours: a retrospective cohort study. Vet J. 2013;**196**:492–498. doi:10.1016/j.tvjl.2012.11.018

[26] London CA, Malpas PB, Wood-Follis SL, Boucher JF, Rusk AW, Rosenberg MP, Henry CJ, Mitchener KL, Klein MK, Hintermeister JG, Bergman PJ, Couto GC, Mauldin GN, Michels GM. Multi-center, placebo-controlled, double-blind, randomized study of oral toceranib phosphate (SU11654), a receptor tyrosine kinase inhibitor, for the treatment of dogs with recurrent (either local or distant) mast cell tumor following surgical excision. Clin Cancer Res. 2009;**15**:3856–3865. doi:10.1158/1078-0432.CCR-08-1860

[27] Marconato L, Zorzan E, Giantin M, Di Palma S, Cancedda S, Dacasto M. Concordance of c-*kit* mutational status in matched primary and metastatic cutaneous canine mast cell tumors at baseline. J Vet Int Med. 2014;**28**:547–553. doi:10.1111/jvim.12266

[28] Amagai Y, Tanaka A, Matsuda A, Jung K, Ohmori K, Matsuda H. Stem cell factor contributes to tumorigenesis of mast cells via an autocrine/paracrine mechanism. J Leukoc Biol. 2013;**93**:245–250. doi:10.1189/jlb.0512245

[29] Amagai Y, Tanaka A, Jung K, Matsuda A, Oida K, Nishikawa S, Jang H, Ishizaka S, Matsuda H. Production of stem cell factor in canine mast cell tumors. Res Vet Sci. 2014;**96**:124–126. doi:10.1016/j.rvsc.2013.10.014

[30] Spector MS, Iossifov I, Kritharis A, He C, Kolitz JE, Lowe SW, Allen SL. Mast-cell leukemia exome sequencing reveals a mutation in the IgE mast-cell receptor beta chain and KIT V654A. Leukemia. 2012;**26**:1422–1425. doi:10.1038/leu.2011.354

[31] Youk J, Koh Y, Kim J, Kim D, Park H, Jung WJ, Ahn K, Yun H, Park I, Sun C. A scientific treatment approach for acute mast cell leukemia: using a strategy based on next-generation sequencing data. Blood Res. 2016;**51**:17–22. doi:10.5045/br.2016.51.1.17

[32] Yoo CB, Jones PA. Epigenetic therapy of cancer: past, present and future. Nat Rev Drug Discov. 2006;**5**:37–50. doi:10.1038/nrd1930

[33] Morimoto C, Tedardi M, Fonseca I, Kimura K, Sanches D, Epiphanio T, Francisco Strefezzi R, Dagli M. Evaluation of the global DNA methylation in canine mast cell tumour samples by immunostaining of 5-methyl cytosine. Vet Comp Oncol. 2016. doi:10.1111/vco.12241

[34] Lin TY, Fenger J, Murahari S, Bear MD, Kulp SK, Wang D, Chen CS, Kisseberth WC, London CA. AR-42, a novel HDAC inhibitor, exhibits biologic activity against malignant mast cell lines via down-regulation of constitutively activated Kit. Blood. 2010;**115**:4217–4225. doi:10.1182/blood-2009-07-231985

[35] Govier SM. Principles of treatment for mast cell tumors. Clin Tech Small Anim Pract. 2003;**18**:103–106. doi:10.1053/svms.2003.36624

[36] Stanclift RM, Gilson SD. Evaluation of neoadjuvant prednisone administration and surgical excision in treatment of cutaneous mast cell tumors in dogs. J Am Vet Med Assoc. 2008;**232**:53–62. doi:10.2460/javma.232.1.53

[37] Matsuda A, Tanaka A, Amagai Y, Ohmori K, Nishikawa S, Xia Y, Karasawa K, Okamoto N, Oida K, Jang H, Matsuda H. Glucocorticoid sensitivity depends on expression levels of

glucocorticoid receptors in canine neoplastic mast cells. Vet Immunol Immunopathol. 2011;**144**:321–328. doi:10.1016/j.vetimm.2011.08.013

[38] McCaw DL, Miller MA, Bergman PJ. Vincristine therapy for mast cell tumors in dogs. J Vet Intern Med. 1997;**11**:375–378

[39] Dobson JM, Cohen S, Gould S. Treatment of canine mast cell tumors with prednisolone and radiotherapy. Vet Comp Oncol. 2004;**2**:132–141. doi:10.1111/j.1476-5810.2004.00048.x

[40] Davies DR, Wyatt KM, Jardine JE, Robertson ID, Irwin PJ. Vinblastine and prednisolone as adjunctive therapy for canine cutaneous mast cell tumors. J Am AnimHosp Assoc. 2004;**40**:124–130. doi:10.5326/0400124

[41] Thamm DH, Turek MM, Vail DM. Outcome and prognostic factors following adjuvant prednisolone/vinblastine chemotherapy for high-risk canine mast cell tumor: 61 cases. J Vet Med Sci. 2006;**68**:581–587.

[42] Rassnick KM, Moore AS, Williams LE, London CA, Kintzer PP, Engler SJ, Cotter SM. Treatment of canine mast cell tumours with CCNU (Lomustine). J Vet Intern Med. 1999;**13**:601–605.

[43] Dobson JM, Scase TJ. Advances in the diagnosis and management of cutaneous mast cell tumours in dogs. J Small Anim Pract. 2007;**48**:424–431. doi:10.1111/j.1748-5827.2007.00366.x

[44] Poirier VJ, Adams WM, Forrest LJ, Green EM, Dubielzig RR, Vail DM. Radiation therapy for incompletely excised grade II canine mast cell tumors. J Am AnimHosp Assoc. 2006;**42**:430–434. doi:10.5326/0420430

[45] Heinrich MC, Griffith DJ, Druker BJ, Wait CL, Ott KA, Zigler AJ. Inhibition of *c-kit* receptor tyrosine kinase activity by STI 571, a selective tyrosine kinase inhibitor. Blood. 2000;**96**:925–932.

[46] Joensuu H, Roberts PJ, Sarlomo-Rikala M, Andersson LC, Tervahartiala P, Tuveson D, Silberman SL, Capdeville R, Dimitrijevic S, Druker B. Effect of the tyrosine kinase inhibitor STI571 in a patient with a metastatic gastrointestinal stromal tumor. N Engl J Med. 2001;**344**:1052–1056.doi:10.1056/NEJM200104053441404

[47] Isotani M, Ishida N, Tominaga M, Tamura K, Yagihara H, Ochi S, Kato R, Kobayashi T, Fujita M, Fujino Y. Effect of tyrosine kinase inhibition by imatinib mesylate on mast cell tumors in dogs. J Vet Int Med. 2008;**22**:985–988.doi:10.1111/j.1939-1676.2008.00132.x

[48] Hahn K, Oglivie G, Rusk T, Devauchelle P, Leblanc A, Legendre A, Powers B, Leventhal P, Kinet J, Palmerini F. Masitinib is safe and effective for the treatment of canine mast cell tumors. J Vet Int Med. 2008;**22**:1301–1309. doi:10.1111/j.1939-1676.2008.0190.x

[49] Hahn KA, Legendre AM, Shaw NG, Phillips B, Ogilvie GK, Prescott DM, Atwater SW, Carreras JK, Lana SE, Ladue T. Evaluation of 12-and 24-month survival rates after treatment with masitinib in dogs with nonresectable mast cell tumors. Am J Vet Res. 2010;**71**:1354–1361. doi:10.2460/ajvr.71.11.1354

[50] Pryer NK, Lee LB, Zadovaskaya R, Yu X, Sukbuntherng J, Cherrington JM, London CA. Proof of target for SU11654: inhibition of KIT phosphorylation in canine mast cell tumors. Clin Cancer Res. 2003;**9**:5729–5734.

[51] Le Cesne A, Blay J, Bui BN, Bouché O, Adenis A, Domont J, Cioffi A, Ray-Coquard I, Lassau N, Bonvalot S. Phase II study of oral masitinib mesylate in imatinib-naive patients with locally advanced or metastatic gastro-intestinal stromal tumour (GIST). Eur J Cancer. 2010;**46**:1344–1351. doi:10.1016/j.ejca.2010.02.014

Section 3

Cardiology

Chapter 5

Chronic Mitral Valve Insufficiency in Dogs: Recent Advances in Diagnosis and Treatment

Sang-Il Suh, Dong-Hyun Han, Seung-Gon Lee,
Yong-Wei Hung, Ran Choi and Changbaig Hyun

Additional information is available at the end of the chapter

http://dx.doi.org/10.5772/64611

Abstract

Chronic mitral valvular insufficiency (CMVI) is the most common acquired heart disease in dogs and is characterized by degenerative valvular changes causing progressive thickening of mitral leaflets and incomplete closure of mitral valve. As the disease progresses, it causes congestive heart failure (CHF) and pulmonary edema if the LA dilation cannot accommodate the volume overload by mitral regurgitation. Therefore, it is the most common cause of cardiac mortality in dogs. This chapter discusses general features of CMVI in dogs focusing on recent advances in diagnosis and treatment.

Keywords: mitral valve , degenerative valve disease , mitral valve insufficiency , heart failure , mitral regurgitation

1. Introduction

Chronic mitral valvular insufficiency (CMVI) is the most common cause of congestive heart failure (CHF) in small breed dogs [1, 2] and is characterized by progressive myxomatous degeneration of the atrioventricular valves [3]. Mitral regurgitation (MR) is the most common sequel to CMVI, which causes volume overload at the left atrial (LA) and left ventricle (LV) and progresses to CHF [4]. Underlying causes of CMVI have yet been identified, although aging and genetic causes were suggested in a certain breed of dogs [5–8] and humans [4]. CMVI is more often seen in old and small breeds of dog. Higher prevalence of CMVI was noticed in Cavalier King Charles Spaniels (CKCS) and Dachshund [9, 10]. CMVI has been noticed in ~50% of 6- to 7-year-old CKCS dogs and ~50% of 10-year-old Dachshund dogs. Those two studies strongly suggested genetic etiology for this disease, although other study found the aging is the major cause for this disease [11]. In these dog breeds, the CMVI is occurred at younger age and progressed more rapidly. List of dog breeds having high prevalence rate for CMVI are Chihuahuas, Malteses, Yorkshire Terriers, Poodles, Papillons, Pekingeses, Miniature Pinschers, Bologneses, Dachshunds, Shih Tzus, Cairn Terriers, Miniature Schnauzers, Bichon Frises, Carvalier King Charles Spaniels, Pugs, West Highland White Terriers, Fox Terriers, Boston Terriers, Welsh Terriers, Whippets, American Cocker Spaniels, Beagles, German Shepherds, and Great Danes [12], although almost one-quarter of dogs over the age of 10 have degenerative changes on mitral valve in any breeds of dog. Higher prevalence rate in Maltese and Shih Tzu has been found in some Asian countries including Korea, Japan, and Taiwan [13].

CMVI is an animal model of human mitral valve prolapse (MVP), which is suggested of polygenic inheritance [14]. Several canine studies also suggested polygenic inheritance for CMVI [6, 8, 15]. Male dogs have higher rate of prevalence almost 1.5 times than female dogs [3]. One study found that the mitral valve was affected in ~60% case of CMVI, and tricuspid valve only was affected in ~10% of CMVI, while both atrioventricular valves (mitral valve and tricuspid valve) were affected in ~30% of CMVI [16].

2. Pathogenesis and clinical signs

Pathological features of canine CMVI are degenerative changes on mitral valve, mitral valve thickening and opacity, several degrees of leaflet retraction, node of valve's end, and lengthened chordae tendineae (**Figure 1**) [17, 18] and are similar to human MVP [19, 20]. Disruption of collagen and deposition of glycosaminoglycans in mitral valve are also common microscopic feature in this disease [11, 19, 21]. Long-standing mitral valvular insufficiency can

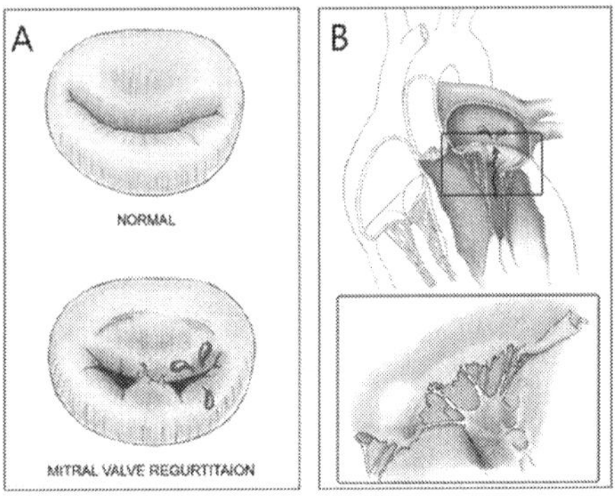

Figure 1. Pathological features of canine chronic mitral valvular insufficiency are degenerative changes on mitral valve, mitral valve thickening and opacity, several degrees of leaflet retraction, node of valve's end and lengthened chordae tendineae. (A) Diagram of normal mitral valve (top) and mitral regurgitation (bottom), (B) diagram of mitral valve insufficiency from chronic degenerative changes on mitral valve leaflets (box).

cause volume overload of both the LA and the LV. Furthermore, the increased end-diastolic volume in LV can cause pressure overload on the LA. Subsequently, the increased pressure in the LA inhibits drainage of blood from the lungs via the pulmonary veins and thus causes pulmonary congestion. If this condition is untreated, it will eventually develop LV dysfunction and CHF [22].

Coughing, especially nocturnal cough, may be the first clinical signs in CMVI. However, in dogs with advanced stage of heart failure (HF), dyspnea such as shortness of breath, difficulty breathing, and orthopnea may be the major signs. Depending on the severity of CMVI, the dog maybe had certain degree of exercise intolerance, lethargy, reduced appetite, and weight loss [23].

3. Diagnosis

3.1. Physical examination

Heart murmur is a major finding in physical examination. Depending on the severity, clinicians can hear various grade of heart murmur before the onset of clinical signs. In early stages of the CMVI, heart murmur can be localized and weak as 1~2/6 scale at left apex. In late stages of the CMVI, heart murmur is gradually radiated and louder and is typically crescendo-decrescendo type systolic regurgitant murmur (**Figure 2A**). Recent study found that the grade of heart murmur was closely related to the severity of CMVI [24].

3.2. Laboratory tests

Common laboratory findings in dogs with CMVI are normal or slightly raise in kidney and/or liver chemistry profiles, probably due to the congestion and poor body perfusion [25]. One recent study evaluated hepatic panel in dogs at different stages of heart failure from CMVI [26].

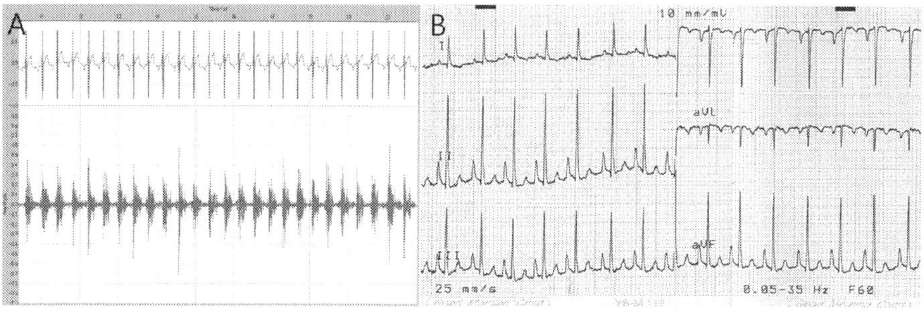

Figure 2. (A) Phonocardiogram in dogs with CMVI. Heart murmur is gradually radiated and louder and is typically crescendo-decrescendo type systolic regurgitant murmur. (B) ECG in dogs with CMVI. P-mitrale (wide P-wave) and wide and tall QRS complexes indicating LA and LV dilation.

Serum levels of ALT and GGT were statistically significantly higher in ISACHC II and III groups ($p < 0.05$), while levels of AST, albumin, cholesterol, and total bilirubin were not significantly differed among groups. Level of NT-proBNP was also significantly higher in ISACHC II and III groups ($p < 0.05$), although the level was not significantly differed in ISACHC I group. There were no correlations between levels of AST, albumin, cholesterol, and total bilirubin to echocardiographic indices. The level of NT-proBNP was correlated with most echocardiographic indices (LA/Ao, LVID/Ao, E-peak, EDVI, $r > 0.7$) and ALT ($r = 0.701$) and GGT ($r = 0.782$). This study revealed the biochemical evidence of hepatic injures in dogs with advanced stage of CMVI [26].

Poor tissue perfusion from CMVI causes pancreatitis in dogs, as indicated by serum pancreatic lipase concentrations. One recent study has evaluated the prevalence of pancreatitis in 62 client-owned dogs consisting of 40 dogs with different stages of heart failure from CMVI and 22 age-matched healthy dogs [27]. Serum canine pancreatic lipase immunoreactivity (cPLI) concentrations were determined by quantitative cPLI test in healthy and CMVI groups in this study. Serum cPLI concentrations were 54.0 μg/L (IQR: 38.0–78.8 μg/L) in control, 55.0 μg/L (IQR: 38.3–88.8 μg/L) in ISACHC I, 115.0 μg/L (IQR: 45.0–179.0 μg/L) in ISACHC II, and 223.0 μg/L (IQR: 119.5–817.5 μg/L) in ISACHC III. Also, close correlation of serum cPLI concentration was found in the left atrial to aorta (LA/Ao) ratio ($r = 0.597$; $P = 0.000$) and the severity of heart failure ($r = 0.530$; $P = 0.000$). This study found that the CMVI is associated with pancreatic injury in congestive heart failure due to the CMVI [27].

Reduction in glomerular filtration rate (GFR) is a common complication in advanced stages of heart failure (HF). The convenient and precise assessment for GFR would be useful for early detection of renal impairment in HF dogs. One recent study has evaluated the reduction in GFR in advanced stages of HF from CMVI, using renal markers including serum cystatin C (Cys-C) and symmetric dimethylarginine (SDMA) concentrations [28]. Forty-three client-owned dogs consisting of 33 dogs with different stages of HF from CMVI and 10 age-matched healthy dogs were enrolled in this study. Serum Cys-C and SDMA concentrations along with other renal (i.e., urea nitrogen and creatinine) and echocardiographic markers were evaluated in healthy and CMVI dogs. Serum Cys-C concentrations were 1.4 ± 0.4 mg/l in control, 2.1 ± 0.9 mg/l in ISACHC I, 2.9 ± 0.8 mg/l in ISACHC II, and 3.6 ± 0.6 mg/l in ISACHC III dogs, whereas serum SDMA concentrations were 8 ± 2 μg/dl in control, 14 ± 3 μg/dl in ISACHC I, 18 ± 6 μg/dl in ISACHC II, and 22 ± 7 μg/dl in ISACHC III dogs. There was close correlation of serum Cys-C and SDMA concentrations with serum creatinine, urea nitrogen, and the severity of HF. This study demonstrated that the GFR was decreased in dogs with CMVI having earlier stages of HF [28].

3.3. Cardiac biomarkers

In recent years, cardiac biomarkers have been developed that are differentiating cardiac and respiratory diseases to evaluate the progress of heart failure in dogs and cats. There are many cardiac biomarkers. The ideal biomarkers should reflect the therapeutic response, the pathophysiology of heart diseases, assist in the early diagnosis of CHF, and be applicable throughout the various phases of the syndrome from before the onset of its clinical manifestations through its end-stage. Cardiac biomarkers have used as diagnostic tools [29], prognostic indicator [30], and monitoring system [31] for CHF.

Troponins are marker of myocardial necrosis and ischemia and found to be closely associated with the severity of heart failure in dogs [32] and cats [33], although it often elevated in many noncardiac disease [34–36]. Natriuretic peptides (NPs) are markers releasing from hemodynamic stress on the heart [37], responded against volume expansion/pressure overload [37]. The plasma concentration of N-terminal prohormone of brain natriuretic peptide (NT-proBNP) is well correlated with severity of heart failure in dogs [38], although the level of NT-proBNP can be affected by noncardiac factors such as body weight and renal function [39]. Cardiopet® proBNP is a commercially available diagnostic test. According to the manufacturer (Idexx, USA), dogs with <900 pmol/L of serum NT-proBNP may not have heart failure, while dogs with 900–1800 pmol/L may have heart failure, but is required further discriminative tests. Dogs with >1800 pmol/L may have higher possibility of heart failure. C-reactive protein (CRP) is an acute-phase reactant protein [40, 41] that is increased in several diseases in dogs [42–47]. Although the level of CRP is increased in dogs with CMVI, the CRP concentration was not related to the presence of CHF or murmur grade [48].

3.4. Electrocardiogram

Major findings on the electrocardiography (ECG) in dogs with CMVI are P mitrale (wide P-wave) and wide and tall QRS complexes indicating LA and LV dilation (**Figure 2B**) [49]. Tachycardia may be occurred either persistently or intermittently as the CMVI progresses [50, 51]. Although atrial fibrillation is often observed in large breed dogs with CMVI, it is rarely found in the small breed dogs. However, if dogs have early stage of CMVI, there will be no abnormal finding on the ECG [23]. The ECG signs indicating myocardial hypoxic damage (i.e., the ST-slurring) can be seen in dogs with advanced stage of heart failure [52].

3.5. Thoracic radiography

Thoracic radiography is the diagnostic test of choice in dogs with CMVI [23]. Enlargement of the LA/LV and pulmonary venous vasculatures is common findings on thoracic radiography (**Figure 3**) [23, 53, 54]. Other radiographic signs indicating left-sided heart failure including the dorsal displacement of trachea, the compression and/or elevation of left main stem bronchus, and the dividing view of left and right stem bronchus can be noticed as the disease progresses (**Figure 3**) [55]. In advanced stage, radiographic signs related to pulmonary edema (i.e., pulmonary venous engorgement, peribronchial pattern, air bronchograms) can be obvious in most cases [23]. Also, when complications with pulmonary hypertension (PHT) are combined, radiographic signs indicating right-sided heart failure (i.e., hepatomegaly, ascites) can be observed [23].

3.6. Echocardiography

The transthoracic echocardiographic examination is noninvasive diagnostic method and can help to identify mitral valvular lesions and to determine the severity of MR. Echocardiography

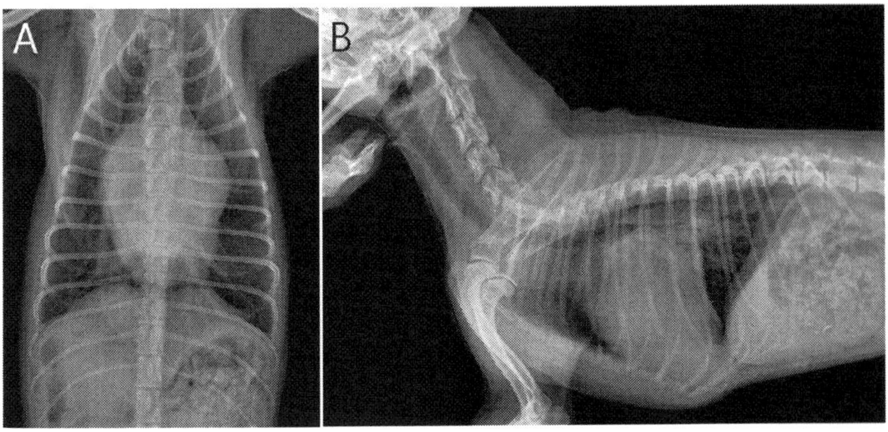

Figure 3. Thoracic radiography in dogs with CMVI. Enlargement of the LA/LV and pulmonary venous vasculatures is common findings on thoracic radiography. Other radiographic signs indicating left-sided heart failure including the dorsal displacement of trachea, the compression and/or elevation of left main stem bronchus, and the dividing view of left and right stem bronchus can be noticed as the disease progresses. (A) Ventrodorsal projection. (B) Right lateral projection.

can also assess its impact on cardiac remodeling, myocardial function, left ventricular filling pressures, and pulmonary arterial pressure [56–61].

Mitral valve lesion can be identified using two-dimensional and M-mode echocardiography. The mitral valve lesions associated with CMVI are small and smooth, creating a club-shaped appearance to the leaflet tips during early stages of the disease, but may become large and irregular during disease progression (**Figure 4B**) [56, 62, 63]. Mitral valve prolapse, which is characterized by one or both leaflets bent back into the left atrial chamber during systole, occurs commonly in dogs with CMVI (**Figure 4A**) [7, 64]. In one recent study, the severity of mitral valve prolapse was significantly correlated with MR severity [64]. Anterior leaflet of mitral valve is more commonly affected than posterior leaflet in dogs [64]. Abnormal excursion [i.e., decreased ejection fraction (EF) slope] and thickening of anterior mitral leaflet can also be detected in M-mode echocardiography (**Figure 5A**).

Ruptured chordae tendineae is also common echocardiographic finding in dogs with CMVI [64]. The mitral valve leaflet is seen pointing back into the left atrium (LA) during systole and bent back on itself with in the left ventricular outflow tract during diastole [65–67]. Chordae tendineae of anterior mitral valve leaflet is more commonly ruptured in dogs [68].

It is clinically important to evaluate severity of MR in dogs with CMVI. Color-flow Doppler imaging (CDI) is widely used for detection and assessment of MR in dogs with CMVI (**Figures 4C**, **D** and **6B**). Maximal area of the regurgitant jet signals to the left atrium area (ARJ/LAA) ratio, which is the maximal ratio of the regurgitant jet area signal to left atrial area, is used in semi-quantification of MR [56, 69, 70]. The ARJ/LAA ratio lesser than 20–30%

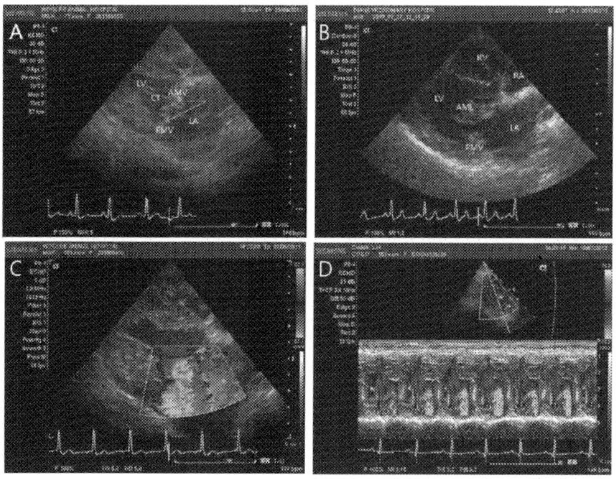

Figure 4. 2D and color Doppler echocardiography in dogs with CMVI. (A) Mitral valve prolapse, which is characterized by one or both leaflets bent back into the left atrial chamber during systole, occurs commonly in dogs with CMVI. The severity of mitral valve prolapse was significantly correlated with MR severity. (B) The mitral valve lesions associated with CMVI are large and irregular in advanced stage of CMVI. Anterior leaflet of mitral valve is more commonly affected than posterior leaflet in dogs. (C) Color-flow Doppler imaging in 2D echocardiography revealed severe regurgitant jets from left ventricle to left atrium during systole and is widely used for detection and assessment of MR in dogs with CMVI. (D) The MR can be also detected in color M-mode echocardiography on the LV short-axis view. LV, left ventricle; CT, chordae tendineae; AMV, anterior mitral valve; PMV, posterior mitral valve; RV, right ventricle; RA, right atrium.

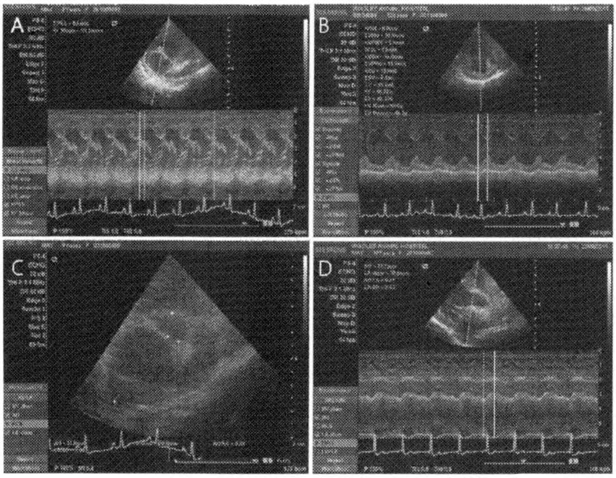

Figure 5. 2D and M-mode echocardiography in dogs with CMVI. (A) Abnormal excursion (decreased EF slope) and thickening of anterior mitral leaflet can be detected in M-mode echocardiography. (B) The eccentric hypertrophy, which is characterized by an increase in end-diastolic left ventricular dimensions (EDV), occurs in dogs with CMVI. (C)–(D) Hemodynamically significant chronic MR can induce volume overload, which subsequently can increase LV and LA volume and can result in LA and LV dilation. The degree of left atrial enlargement that is assessed by the left atrium to aorta (LA/Ao) ratio in 2D and M-mode echocardiography and is closely correlated with the severity of heart failure.

is indicative of mild MR. The ratio involves between 20 and 30, 70% is indicative of moderate MR, and >70% is indicative of severe MR [56, 69–71].

The vena contracta is the measurement of smallest mitral regurgitant jet width through the valve and is correlated with MR severity [71]. Measuring the vena contracta uses parasternal long-axis views that identify the vena contracta perpendicular to the sound plane. Very few data are available regarding vena contracta associated with dogs with CMVI [71]. In human study, a correlation was found between vena contracta and MR severity [72–74]. The proximal isovelocity surface area (PISA) method can be used in echocardiography to estimate the area of flow acceleration and convergence proximal to the mitral valve as the regurgitant jet approaches the orifice [75]. Left apical four-chamber views confirming the mitral regurgitant jet are generally recommended for measurement. Regurgitant orifice area, regurgitant fraction, and volume can be measured by this method. In study performed on dogs with severe MR, mitral regurgitant fraction calculated using PISA method was significantly correlated with MR severity [76].

Hemodynamically significant chronic MR can induce volume overload, which subsequently can increase LV and LA volume and can result in LA and LV dilation [63]. The degree of left atrial enlargement that is assessed by the left atrium to aorta (LA/Ao) ratio is closely correlated with the severity of heart failure (**Figure 5C** and **D**). Both two-dimensional mode and M-mode echocardiography should be used in order to determine left atrial enlargement. Other indirect signs of high LA pressures in dogs with CMVI are left atrial rupture or acquired atrial

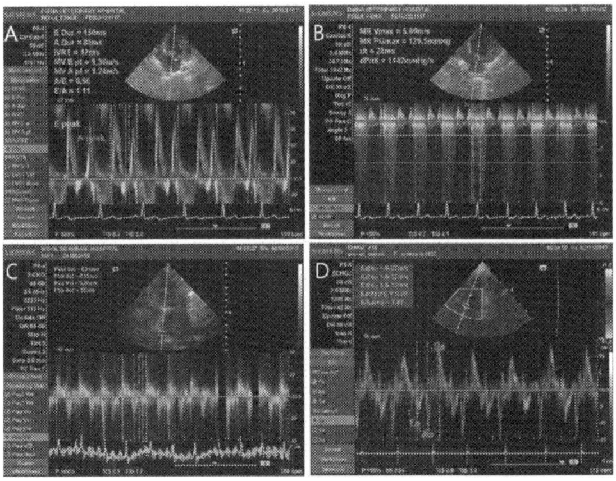

Figure 6. Pulse and continuous Doppler and tissue Doppler echocardiography in dogs with CMVI. (A) The transmitral flow profile consists of E and A and is affected by the pressure gradient between the LA and LV. Elevated E represents increased LA pressure and a worsening of heart failure. (B) Continuous Doppler echocardiography is useful to detect MR in dogs with CMVI. However, the degree of MR is not correlated with the severity of CMVI. (C) Pulse Doppler echocardiography in pulmonary venous flow is also useful to assess the progression of CMVI. The presence of pulmonary venous flow at atrial systole (PVa) indicates high LA pressure noticed in advanced stage of CMVI. (D) The early mitral inflow velocity to early mitral annular tissue velocity (E:Ea) ratio can be used to assess LV diastolic function. The E:E′ ratio is significantly correlated with left ventricular filling pressures.

septal defect secondary to atrial septal rupture [77–79]. Due to volume retention, remodeling, and elevations in LA and pulmonary venous pressures, left ventricular volume overload can occur concomitantly with MR worsening [80]. The eccentric hypertrophy, which is characterized by an increase in end-diastolic left ventricular dimensions (EDV), occurs in dogs with CMVI (**Figure 5B**) [81]. The diastolic left ventricular volume and diameters should be assessed by both M-mode and two-dimensional echocardiography.

The common indices of left ventricular systolic myocardial function are ejection fraction (EF) and fractional shortening (FS). The EF is defined by the percent of the end-diastolic volume ejected from left ventricle (LV) with each heart beat. The FS is a measure of the percent change in the dimension from end diastole to end systole [82]. FS is dependent on preload, afterload, and myocardial contractility. Left ventricular systolic myocardial dysfunction is traditionally characterized by reduced EF and FS. However, FS should be increased in dogs with CMVI because of elevated preload and reduced afterload, and hyperdynamic ventricular contraction.

Systolic myocardial dysfunction may be identified by end-systolic left ventricular dimensions, such as end-systolic diameter, end-systolic volume, and end-systolic volume indexed to body surface area (ESVI) [60, 61, 83]. Increased end-systolic left ventricular dimensions are consistently referred to impaired left ventricular systolic function. There are three ultrasound methods, including the Teichholz method, the monoplane Simpson's derived method of discs, and the length-area method, to measure ESVI. Because the Teichholz method tends to overestimate ESVI, other methods except Teichholz method may be recommended in dogs with CMVI [61].

Spectral Doppler methods, including pulsed wave Doppler (PWD) and continuous wave Doppler (CWD), can identify regurgitant jet, transmitral flow in dogs with CMVI (**Figure 6A** and **B**). PWD may be used to record transmitral flow profile. The sample gate should be located in the tip of the leaflet. The transmitral flow profile looks like "M-shaped" and consists of E (peak early transmitral flow velocity) and A (peak late transmitral flow velocity) related to early filling and atrial contraction, respectively [84]. Diastolic function and left ventricular filling pressures can be assessed using PWD method [84]. CWD is generally used for assessing severity of MR and thus provides information on LA pressure, preload, and systemic arterial pressure. CWD can be used to identify elevated LA pressure, left ventricular systolic and diastolic dysfunction in dogs with CMVI [60]. The presence of pulmonary venous flow at atrial systole (PVa) indicates high LA pressure and is commonly noticed in advanced stage of CMVI (**Figure 6C**). Continuous Doppler echocardiography is useful to detect MR in dogs with CMVI (**Figure 6B**). However, the degree of MR is not correlated with the severity of CMVI.

Advanced echocardiographic techniques, including tissue Doppler image (TDI), strain and strain rate imaging, and speckle tracking echocardiography (STE), are recently developed to assess myocardial abnormalities. TDI measures the myocardial velocities to quantify myocardial abnormalities. Systolic and diastolic myocardial abnormalities can be detected by TDI. TDI can be used in PWD and CDI. Myocardial velocity profile is characterized by a S wave, an E wave, and an A wave related to systolic myocardial velocity, early diastolic myocardial relaxation velocity, and late diastolic myocardial relaxation velocity, respectively. Strain and strain rate imaging are TDI-based measurement and represent regional myocardial deformation and deformation rate, respectively [85, 86]. The two-dimensional STE is recently available and can be used to assess regional myocardial function. The STE is created by irregularities in reflected ultrasound from neighboring structures [87]. Very few data are available regarding advanced echocardiographic techniques data associated with canine heart diseases. TDI provides myocardial and annular velocity. Unlike Doppler patterns of mitral inflow, TDI assessment of diastolic function is relatively load-independent. The early mitral inflow velocity to early mitral annular tissue velocity (E:Ea) can be used to assess LV diastolic function (**Figure 6D**). The E:Ea ratio is significantly correlated with left ventricular filling pressures [88, 89].

Because most dogs affected by CMVI are older, progression of CMVI can lead to diastolic dysfunction, with time. Diastolic dysfunction is characterized by increased resistance to filling and increased left ventricular filling pressure secondary to decreased compliance and impaired relaxation [90, 91]. The assessment of left ventricular diastolic function is difficult to undertake in the dogs with CMVI. Diastolic function can be assessed using several parameters, including isovolumetric relaxation time (IVRT), transmitral flow velocities, and myocardial velocities.

Elevated LA pressure caused by MR and volume overload was found in dogs with moderate-to-severe CMVI [92, 93]. For the noninvasive assessment of LA pressure, the IVRT can be used in volume overload model [93]. IVRT is the time that elapses from aortic valve closure to mitral valve opening. In recent studies, the duration of IVRT and the ratio of E to IVRT were used in the diagnosis of elevated LA pressure [94, 95]. Decrease in IVRT is indicative for increase in LA pressure.

The left ventricular diastolic function can be assessed by Doppler patterns of mitral inflow. The transmitral flow profile consists of E and A and is affected by the pressure gradient between the LA and LV. Elevated E represents increased LA pressure and a worsening of heart failure (**Figure 6A**) [5]. If diastolic function is normal, E is greater than A. In early diastolic dysfunction, a reversal of E and A can be occurred, as left ventricular compliance decreases. Further worsening of diastolic function leads to pseudonormalization associated with increased LA pressure. Because mitral inflow velocities are load-dependent, the use of transmitral flow profile to assess diastolic function remains limited.

One recent study has evaluated the diagnostic value of left atrial volume index (LAVi) and the ratio of early filling to early diastolic mitral annular velocity (E/Ea) on the progression of heart failure in 51 dogs with CMVI and body weight matched 18 healthy control dogs, along with other known echocardiographic markers [96]. The LAVi and E/Ea were well correlated with the severity of heart failure in this study group. Based on the receiver-operating characteristic analysis on echocardiographic variables, the echocardiographic indications for advanced heart failure in this study were left atrium to aorta ratio (LA:Ao) >2.0, left ventricular diastolic dimension to aorta ratio (LVIDd:Ao) >2.4, end-diastolic volume index (EDVI) >100 ml/m^2, transmitral E-peak >1.2 m/s, E/Ea >9.0 and LAVi 49 ml/m^2, while indications for healthy or dogs with no signs of cardiac enlargement were LA:Ao <1.3, LVIDd:Ao <1.7, EDVI <45 ml/m^2, E-peak <0.65 m/s, E/Ea <6.0 and LAVi <15 ml/m^2 in dogs with CMVI.

4. Treatment

Before initiating treatment for dogs with CMVI, proper staging of heart failure in each affected dogs is required. Two classification systems can be applied in dogs: International Small Animal Cardiac Health Council System (ISACHC) and American College of Veterinary Internal Medicine (ACVIM) classification (**Table 1**).

Basic strategies for treating CMVI are (1) to lessen cardiac workload, (2) to improve clinical conditions from CHF, (3) to retard cardiac remodeling from neurohormonal response from heart failure, and (4) to reduce complications from heart failure. Several therapeutic strategies have been recommended in the veterinary literatures [18, 97–101]. In 2009, the ACVIM released an expert consensus statement and provided therapeutic guideline for dogs with

ISACHC		ACVIM	
		A	Patient at risk of developing heart disease in the future, e.g., patient from breed with high predisposition for cardiac disease
Ia	Asymptomatic: no evidence of compensation for underlying heart disease (no volume overload or pressure overload detected radiographically or echocardiographically)	B1	Asymptomatic patients with evidence of structural heart disease, e.g., presence of murmur: with no evidence of cardiac remodeling (radiographically or echocardiographically)
Ib	Asymptomatic: clinical signs of compensation for underlying heart disease (volume overload or pressure overload detected radiographically or echocardiographically)	B2	Asymptomatic patients with evidence of structural heart disease, e.g., presence of murmur: with evidence of cardiac remodeling
II	Mild-to-moderate heart failure with clinical signs at rest or with mild exercise. Treatment required	C	Patients with clinical signs of congestive heart failure (either past or present) Stage C1 (stabilized CHF) Stage C2 (mild to moderate CHF) Stage C3 (severe and/or life threatening CHF)
IIIa	Advanced heart failure; clinical signs of severe congestive heart failure: home treatment possible		
IIIb	Advanced heart failure; clinical signs of severe congestive heart failure: requires hospitalization	D	Refractory heart failure. Patients showing clinical signs in spite of standard treatment for congestive heart failure

Table 1. International Small Animal Cardiac Health Council System (ISACHC) and American College of Veterinary Internal Medicine (ACVIM) classification in dogs with heart failure.

CMVI [3]. In practice, the first-line medications for heart failure in dogs with CMVI should include furosemide, pimobendan, and angiotensin-converting enzyme (ACE) inhibitor. The route and dose of furosemide administration should be adjusted based on the degree of respiratory distress and disability. Monitoring renal function is necessary for every dogs, especially before and 3–5 days after initiation and adjustment of furosemide and an ACE inhibitor (ACEI). Either surgical replacement or valvuloplasty of damaged mitral valve has been successfully applied in dogs. Furthermore, experimental prosthetic devices for treating CMVI in dogs are under development and evaluation [102, 103].

4.1. Guidelines for long-term management of CMVI

Stage A (risk for heart failure): Certain breeds of dogs with genetic etiologies, family history of heart disease, a breed predisposition, or concurrent systemic disease with cardiovascular implications (e.g., Cavalier King Charles Spaniels) may have high risk of heart diseases. In these dogs, periodical monitoring for heart diseases is necessary, although no specific therapy is required before the evidence of heart diseases is detectible, according to recent guideline from ACVIM [3]. Dogs used for breeding should be removed from the breeding program if CMVI is present in earlier life. It should be recommended to the dog's owner for periodic cardiac examinations. It is also recommended to manage predisposing condition and to manage systemic hypertension, if present. No dietary sodium modifications are necessary in this stage.

Stage B1 (heart disease is present: no symptoms, no obvious chamber enlargement): It is better to inform the owner clinical signs related to CHF (tachypnea, dyspnea, coughing) as early as possible. Periodic reevaluation for signs of disease progression and complications is necessary. For patients with CMVI, there is no evidence indicating that there is any beneficial effect of using an ACE inhibitor (ACEI) or pimobendan at this stage.

Stage B2 (heart disease is present: no symptoms, cardiomegaly present): It is generally recommended the use of ACEI (enalapril 0.5 mg/kg PO sid to bid; benazepril 0.5 mg/kg PO sid) and highly palatable mildly sodium-restricted diet. Some cardiologists suggested the use of spironolactone 1 mg/kg PO bid for possible aldactone escape. More detailed guideline can be found in the section "Guideline for asymptomatic dogs with CMVI."

Stage C1 (stabilized CHF): If dogs had historical signs of congestive heart failure (CHF), but had no symptoms currently, it is important to keep clinical signs stabilized. Drugs for routine use are furosemide (mandatory) along with ACEI, and/or pimobendan. Drugs for selected patients are spironolactone, digoxin, thiazide, amlodipine/hydralazine, or other vasodilator. In this stage of dogs, excessive sodium intake, beta-blockers, corticosteroid, and intravenous (IV) fluid should be avoided, if possible (unless required for concurrent disease). If IV fluid is given to this dog, it requires careful monitoring of the respiratory rate trend.

Stage C2 (mild-to-moderate CHF): Therapeutic goals are to eliminate pulmonary edema or effusions, to improve hemodynamics, and to modulate neurohormonal activation. In dogs with CMVI, drugs include furosemide (1–2 mg/kg PO bid), ACEI (enalapril 0.5 mg/kg PO bid), and pimobendan (0.25 mg/kg PO bid). Digoxin may be beneficial, if atrial fibrillation is present. Beta-blockers should not be introduced firstly, unless the dog is being medicated. In this stage of dogs, excessive sodium intake, beta-blockers, corticosteroid, and intravenous (IV) fluid should be avoided, if possible (unless required for concurrent disease). If IV fluid is given to this dog, it requires careful monitoring of the respiratory rate trend.

Stage C3 (severe and/or life threatening CHF): Therapeutic goals are to treat hypoxemia, to increase cardiac output, and to stabilize the patient in hospital with intravenous drugs. Drugs for routine use in stage C2 are needed with oxygen supplementation (depending on dog's condition) and high dose of furosemide (2–8 mg/kg IV; repeat injections every 1–2 h if there is no improvement in respiratory rate) with nitrate therapy (e.g., nitroglycerin patch/cream, sodium nitroprusside, isosorbide dinitrate). As the respiratory rate decreases, the dosage and frequency of administration are reduced to the lowest dose effective in controlling the pulmonary edema. Renal function should be kept monitoring.

Stage D (refractory, chronic CHF): Drugs for routine use in stage C3 are needed with increased dose/frequency of pimobendan (up to 0.7 mg/kg, PO, tid), supplementation of spironolactone (1–2 mg/kg PO, bid) and hydrochlorothiazide (1–2 mg/kg, PO, bid), subcutaneous furosemide, repeated centesis for effusions, digoxin or other antiarrhythmic drugs if needed, and very low sodium intake. Triple diuretics (furosemide, spironolactone, hydrochlorothiazide) can reduce the dose of furosemide required to control the patient's congestive signs. In CMVI, it can be considered for additional amlodipine (0.05 mg/kg PO, sid, then 0.1 mg/kg PO, with blood pressure monitoring) if blood pressure is normally preserved.

4.2. Guidelines for short-term management (acute pulmonary edema) of CMVI

Stage Ca (acute heart failure requiring hospitalization): The goals of therapy are to relieve the severe pulmonary edema. For dogs with pulmonary edema from acute pulmonary edema, the therapy should be directed (1) to reduce the circulating blood volume by either/both aggressive and immediate diuretic therapy (e.g., furosemide IV or CRI) and/or phlebotomy (10 ml/kg), (2) to reduce the venous return to the cardiac chambers [e.g., topical 2% nitroglycerin cream, intravenous acepromazine, intravenous sodium nitroprusside (SNP) CRI], (3) to increase oxygen saturation (e.g., oxygen tent or nasal oxygen), and (4) to strengthen myocardial systolic function (e.g., intravenous dobutamine 5–15 μg/kg/min CRI).

Stage Da (refractory heart failure requiring hospitalization): Aggressive furosemide therapy [4 mg/kg IV followed by repeat injections every 4 h or 4 mg/kg IV followed by CRI (0.2–1 mg/kg/hr for 8–12 h)] should be initiated as early as possible, till respiratory rate has fallen by 50%. Intravenous sodium nitroprusside (SNP) therapy along with furosemide would be beneficial to stabilize CHF dogs. To achieve this therapeutic goal, intravenous infusion of SNP should be administered at 2 μg/kg/min and then increased by increments of 1 μg/kg/min every 30 min (maximum dose should not be over 6 μg/kg/min) to reach desirable therapeutic effect, if mean and systolic blood pressure of dogs maintain above 75 and 90 mmHg, respectively. Intravenous inotropic support using dobutamine is often also required. Intravenous dobutamine infusion should be started at 5 μg/kg/min and then increased by 2.5 μg/kg/min every 4 h to reach therapeutic effect (a maximal dose of 15 μg/kg/min). The dose rate should be adjusted by mean heart rate (HR) of dogs. The infusion rate should be reduced, if the heart rate increases by 10% or rises over 180 bpm. It is also recommended to supply oxygen by tent, cage, mask, and neck collar or even mechanical ventilation. Clinicians should relieve dyspnea/discomfort via appropriate humidity, environmental temperature, and body positioning during oxygen supplementation.

4.3. Guideline for asymptomatic dogs with CMVI

Several studies have evaluated what cardiac medications can retard the progression of heart failure and can be more effective in asymptomatic HF dogs [18, 104], although most monotherapy was not able to achieve these goals, to date. One recent study has evaluated the outcome of dogs with preclinical cardiomyopathy with atrial fibrillation after either pimobendan monotherapy or benazepril monotherapy, and has found that pimobendan monotherapy provided significantly better outcome (i.e., prolonged time to onset of HF or reduced incidence of sudden death [105]). Unfortunately, several studies failed to find beneficial effects on survival and onset of HF in asymptomatic dogs with various heart diseases after the long-term administration of ACEI including enalapril [99, 105]. One recent small pilot study in dogs with asymptomatic HF found modest evidence of beneficial effect on retarding the onset of clinical HF after pimobendan and enalapril dual therapy [106]. One other recent study in asymptomatic dogs with CMVI has also found echocardiographic evidences on improvement of cardiac performance (i.e., increased %LVEF and decreased ESVI) for the first few months after pimobendan monotherapy [107], although this effect did not last to the end of test period (6 months). One recent study on preclinical CMVI dogs after long-term treatment of enalapril has found long-term administration of enalapril could significantly delay onset of HF and the endpoint of HF-all-cause death [104], although the other study in asymptomatic Cavalier King Charles Spaniels with CMVI has failed to find this beneficial effect [99].

4.4. New therapeutic agents in dogs with CMVI

Isosorbide dinitrate (ISDN) is a moderate- to long-acting organic nitrate, and its venodilatory effects may help reduce preload and hence pulmonary edema. In humans, ISDN is used for treating or preventing angina, treating esophageal spasm and achalasia [108, 109]. In addition, it is widely used for CHF outpatients as an adjunctive treatment in CHF [110, 111]. In dogs, it occasionally used to adjuvant agent for management of chronic heart failure or in combination with an arteriolar dilator for patients unable to tolerate an ACEI [112]. However, there is limited experience in using this drug in veterinary medicine, and adverse effect is not well known. In humans, the most common adverse effects are headache and postural hypotension. Tachycardia, restlessness, or gastrointestinal effects are not uncommon. There have been rare cases of patients who are hypersensitive to organic nitrates. One recent study has evaluated the efficacy of ISDN for treating advanced stage CHF due to CMVI [113]. Twenty dogs with CMVI were enrolled in this study. All dogs were administered sustained-release ISDN (1 mg/kg, q12hr, PO) along with conventional cardiac medication. Changes in systolic blood pressure (SAP), heart rate (HR), and echocardiographic indices indicating the progression of CHF were evaluated at 7, 15, 30, and 60 days after the administration of ISDN. Significant improvements in echocardiographic indices were found at 7, 15, 30, and 60 days after the administration of ISDN, although the Systolic arterial pressure (SAP) was slightly decreased and the HR

http://dx.doi.org/10.5772/64611

was slightly increased. This study suggested that ISDN could effectively reduce the cardiac preload and thus improve cardiac performance in dogs with advanced heart failure [113].

Angiotensin receptor blockers (ARBs) inhibit type I angiotensin II (AT1) receptor distributed in blood vessels and heart, and thus exert similar pharmacological action of ACEI. Because the ARBs only block type I receptor, they can reduce risk of renal injury from full inhibition of ACE [114]. Therefore, it can use for treating dogs with CHF, when the ACEIs cause renal azotemia [115]. However, the application of these agents on veterinary medicine is limited due to lack of studies related to ARBs in dogs. The common ARBs in veterinary fields are candesartan, losartan, valsartan, and telmisartan.

Pimobendan is a benzimidazole-pyridazinone drug which is used commonly for treating various heart diseases in dogs including CHF. It acts through calcium sensitization and inhibition of phosphodiesterase III [116, 117]. Pimobendan has vasorelaxation effect by inhibition of phosphodiesterase III and positive inotropic effect through calcium sensitization in myocardial sarcomere [118, 119]. Pimobendan can improve myocardial contractility without increasing the risk of arrhythmia unlike digitalis, because this drug does not require oxygen consumption of myocardium [100, 120]. Pimobendan can effectively decrease afterload and peripheral vessel resistance by relaxing vascular smooth muscle through inhibiting a vasoconstriction factor like PDE III [121, 122]. Pimobendan can also delay inflammatory response of myocardium and can improve myocardial contraction by weakening revelation of inflammation precursor and nitric oxide synthesis [123, 124]. Pimobendan can increase sinus rate through rising of blood volume in normal dogs, although it rarely causes arrhythmia unlike digitalis [125, 126]. Therefore, those pharmacological effects are very useful for control clinical signs associated with CHF in dogs and have been well documented in veterinary literatures [127, 128].

5. Complications and prognosis

There are some complications due to heart failure from CMVI such as ruptured chordae tendineae (RCT), pulmonary hypertension (PHT), acute exacerbation of pulmonary congestion, LA rupture, and cardio-renal syndrome (CRS) caused by forward heart failure from CMVI [23, 129].

5.1. Deterioration of cardiac disease

RCT can cause acute exacerbation of pulmonary congestion and edema in dogs with CMVI. In particular, if the first-order chordae attached to the septal leaflet is ruptured, the clinical signs tend to more rapidly aggravate from acute volume overload and fulminant pulmonary edema [68], although the RCT in different mitral leaflet may not cause significant clinical signs. In dogs with significant RCT, marked increase in LA and pulmonary venous pressures can lead to acute pulmonary edema, pulmonary artery hypertension, and right-sided heart failure [22, 130]. Therefore, these dogs usually require intensive care to stabilize the condition along with the standard therapy for CMVI.

Right-sided heart failure is common, especially in dogs with long-standing history of CMVI. Right-sided heart failure can be occurred by either/both concurrent chronic tricuspid insufficiency from myxomatous degeneration and/or the PHT from LA volume and pressure overload. Dogs with marked PHT generally show marked exercise intolerance with signs of weakness or collapse. Signs related to right-sided heart failure (e.g., ascites, pleural effusion, hepatic and splenic congestion, and distention of the jugular veins with abnormal pulsations) can be noticed in physical examination. The presence and degree of PHT can be accurately assessed by Doppler echocardiography. Oxygen supplementation and pulmonary arterial vasodilator (e.g., sildenafil) are helpful to lessen clinical signs in dogs [131, 132].

Tachyarrhythmias are more commonly occurred in dogs having an enlarged LA. Common tachyarrhythmias in dogs with CMVI are supraventricular premature beats, atrial fibrillation, and supraventricular tachycardia. If the tachyarrhythmia has ventricular rate >180 bpm, it can cause hemodynamically significant change in dogs with CMVI and cause an acute onset of pulmonary edema. This condition is more often seen in dogs with long-standing CMVI. Therapeutic goals for these dogs are directed to relieve the pulmonary edema along with the reduction in heart rate to an acceptable rate for improving cardiac output.

Left atrial rupture and cardiac tamponade can be occurred by marked dilation of LA in dogs with CMVI, because the LA becomes thin walled and more vulnerable to increase in pressure. One study found that endocardial splitting is more common in dogs with long-standing CMVI [133]. Long-standing and marked LA volume and pressure overload can progress to rupture of the LA, subsequently with the acute onset of hemopericardium, cardiac tamponade, and sudden death. Acute development of ascites, collapse, or marked exercise intolerance can be signs for sudden development of LA rupture and cardiac tamponade. Echocardiography is necessary for confirming the presence of significant pericardial effusion. Although immediate pericardiocentesis may be helpful to alleviate clinical signs, the prognosis is usually poor.

5.2. Cardio-renal syndrome

Cardio-renal syndrome (CRS) is a clinical syndrome broadly in which dysfunctional hearts and dysfunctional kidneys can "initiate and perpetuate disease in the other organ though common hemodynamic, neurohormonal, and immunological/biochemical feedback pathways" [134]. General definition of CRS is a pathophysiologic disorder of the heart and kidneys, whereby acute or chronic dysfunction in one organ may induce acute or chronic dysfunction in the other organ. According to human medical literature [134], the CRS is largely divided into five types: (1) Type I (acute CRS) is acute kidney injury induced by acute heart failure (e.g., acute cardiogenic shock or acutely decompensated CHF), (2) Type II (chronic CRS) is permanent and progressive chronic kidney disease induced by chronic heart failure (e.g., chronic abnormalities in cardiac function), (3) Type III (acute reno-cardiac syndrome, RCS) is acute heart failure induced by acute kidney diseases (e.g., acute kidney ischemia or glomerulonephritis), (4) Type IV (chronic RCS) is chronic heart failure induced by chronic kidney disease (e.g., chronic glomerular or interstitial disease), and (5) Type V (secondary CRS) is heart and renal failure induced by systemic diseases (e.g., diabetes mellitus, sepsis). Major mechanisms of CRS include renal hypoperfusion directly resulting from a decreased cardiac output and neurohormone-mediated renal damage as hypertensive nephropathy via activation of the renin-angiotensin-aldosterone system (RAAS) among others.

In veterinary study, the prevalence of azotemia is high in dogs with CMVI and increases with the severity of the heart failure and azotemia is associated with a decrease in GFR [135]. Azotemia and renal impairment increase with the severity of CHF and are frequent findings in dogs with CMVI [129]. One retrospective study of 33 dogs with CMVI demonstrated that the prevalence of azotemia (defined as abnormally elevated serum levels of Cys-C, SDMA, and creatinine) was increased in dogs with CMVI [28]. Azotemia and renal impairment increase with the severity of HF and are frequent findings in dogs with CMVI [129]. Keys for successful management of CRS are: (1) try to decrease the dosage of furosemide if azotemia was worsen during the CHF treatment, (2) increase water intake, (3) consider IV fluid, if patients have clinical sings of azotemia (e.g., 2.5% dex + 0.45% saline, 5% dextrose; 30–40 ml/kg/day), and (4) monitor patient's condition regularly to maintain proper dose of furosemide/ACEI.

5.3. Impaired function of digestive system

Impaired function of digestive system is associated with malassimilation (i.e., maldigestion and malabsorption) induced weight loss [136]. The weight loss is a major clinical finding in certain degenerative diseases including CMVI [23], hepatobiliary disease [137], and pancreatitis [138]. Aging is involved in the pathogenesis of CMVI in dogs [6–8, 139] and can induce several anatomical changes and involve in progressive deterioration of the vital physiological functions. The organ congestion and poor body perfusion can be occurred by heart failure [25], and these can lead to organ damage (i.e., pancreas, liver, intestine) and dysfunction (i.e., maldigestion and malabsorption, hepatomegaly, ascites) [23, 140–144]. Therefore, pancreatic dysfunction associated with heart failure can be occurred in dogs with advanced stage of heart diseases. Ischemia can induce acute/chronic pancreatitis and is one of the important etiologies of acute pancreatitis in human. Several mechanisms are involved in pathogenesis of pancreatitis, such as hemorrhage or hypotension, mesenteric macro-vessel occlusion, post-transplantation pancreatitis, cardiopulmonary bypass. Different causes of ischemia can lead to a hypoperfusion of the pancreas with a consecutive induction of an inflammatory response. Diagnosis of pancreatitis is straightforward with in-house diagnostic test kit (SNAP® cPLI™), which has 95% correlation on sensitivity to the reference laboratory. One recent study found that CMVI is associated with pancreatic injury in congestive heart failure caused by CMVI. Therefore, periodic monitoring on cPLI could be useful in monitoring dogs in heart failure [27].

Oral cavity is one of the blood-rich organs. Hypoxia can induce many dental problems including dental tartar and periodontitis, which is requiring general anesthesia for treatment. We developed anesthetic protocol for dogs with cardiac diseases [145]. Our study was designed to evaluate the effects on cardiovascular system in dogs using anesthetic combination of alfaxalone, butorphanol, and midazolam. Compared to the baseline value (before anesthesia), all cardiac indices were decreased, although only heat rate and aortic blood pressure were statistically significantly decreased ($p < 0.05$). However, the cardiac depression was minimal and transient by this combination of anesthetic agents [145].

Cardiac cachexia is generally seen in dogs with history of long-standing CMVI [23, 99]. Once heart failure develops, an important indicator of a worsening condition is the occurrence of cardiac cachexia, which is unintentional rapid weight loss (a loss of at least 7.5% of normal weight within 6 months).

5.4. Prognosis

Many dogs with CMVI may live for years before developing any symptoms. Prognosis for dogs with CMVI is greatly depending on the severity of heart failure and duration and quality of medical therapy and patient monitoring. Generally, the average survival time of dogs with CMVI is ~3 years in dogs with ISACHC I stage heart failure, while 1–3 years in dogs with ISACHC II stage heart failure and ~6–12 months in dogs with ISACHC III stage of heart failure, respectively [13]. There are several prognostic indicators for dogs with CMVI. The degree of exercise intolerance [146], degree of cardiomegaly [53], degree of LA/LV enlargement [7, 147], and certain ECG indices (e.g., the degree of tachycardia or vagus tone index) [51] were closely related to the prognosis. Furthermore, certain echocardiographic indices (e.g., LA/Ao ratio, E/A wave ratio, EDVI) were found to be a good prognostic indicator in dogs with CMVI [84, 148]. Weight loss (e.g., cachexia) and degree of azotemia by reduced glomerular filtration rate (GFR) are indicators for worsening clinical signs.

Author details

Sang-Il Suh[1], Dong-Hyun Han[1], Seung-Gon Lee[2], Yong-Wei Hung[3], Ran Choi[4] and Changbaig Hyun[1]*

*Address all correspondence to: hyun5188@kangwon.ac.kr

1 Section of Small Animal Internal Medicine, College of Veterinary Medicine, Kangwon National University, Chuncheon, Korea

2 Seoul Animal Heart Hospital, Seoul, Korea

3 Cardiospecial Veterinary Hospital, Taipei, Taiwan

4 Cardiology Section, Dasom Animal Medical Center, Busan, Korea

References

[1] Buchanan JW: Prevalence of cardiovascular disorders. In: Fox PR, Sisson D, Moise NS, editors. Textbook of canine and feline cardiology, 2nd ed. Philadelphia: WB Saunders; 1999. p. 457–470.

[2] Häggström J, Pedersen DH, Kvart C: New insights into degenerative mitral valve disease in dogs. Vet Clin North Am Small Anim Pract. 2004; **34**: 1209–1226. doi:10.1016/j.cvsm.2004.05.002

[3] Atkins C, Bonagura J, Ettinger S, Fox P, Gordon S, Häggström J, Hamlin R, Keene B, Luis-Fuentes V, Stepien R: Guidelines for the diagnosis and treatment of canine chronic valvular heart disease. J Vet Intern Med. 2009; **23**: 1142–1150. doi:10.1111/j.1939-1676.2009.0392.x

[4] Fox PR: Pathology of myxomatous mitral valve disease in the dog. J Vet Cardiol. 2012; **14**: 103–126. doi:10.1016/j.jvc.2012.02.001

[5] Borgarelli M, Savarino P, Crosara S, Santilli RA, Chiavegato D, Poggi M, Bellino C, La Rosa G, Zanatta R, Haggstrom J, Tarducci A: Survival characteristics and prognostic variables of dogs with mitral regurgitation attributable to myxomatous valve disease. J Vet Intern Med. 2008; **22**: 120–128. doi:10.1111/j.1939-1676.2007.0008.x

[6] Olsen LH, Fredholm M, Pedersen HD: Epidemiology and inheritance of mitral valve prolapse in dachshunds. J Vet Intern Med. 1999; **13**: 448–456. doi:10.1111/j.1939-1676.1999.tb01462.x

[7] Pedersen D, Lorentzen K, Kristensen B: Echocardiographic mitral valve prolapse in cavalier king charles spaniels: epidemiology and prognostic significance for regurgitation. Vet Rec. 1999; **144**: 315–320. doi:10.1136/vr.144.12.315

[8] Swenson L, Häggström J, Kvart C, Juneja R: Relationship between parental cardiac status in cavalier king charles spaniels and prevalence and severity of chronic valvular disease in offspring. J Am Vet Med Assoc. 1996; **208**: 2009–2012.

[9] Häggström J, Hansson K, Kvart C, Swenson L: Chronic valvular disease in the cavalier king charles spaniel in Sweden. Vet Rec. 1992; **131**: 549–553. doi:10.1136/vr.131.24.549

[10] Pedersen HD, Nørby B, Lorentzen KA: Echocardiographic study of mitral valve prolapse in dachshunds. Zentralbl Veterinarmed A. 1996; **43**: 103–110. doi:10.1111/j.1439-0442.1996.tb00433.x

[11] Buchanan JW: Chronic valvular disease (endocardiosis) in dogs. Adv Vet Sci Comp Med. 1977; **21**: 75–106.

[12] Parker HG, Kilroy-Glynn P: Myxomatous mitral valve disease in dogs: does size matter? J Vet Cardiol. 2012; **14**: 19–29. doi:10.1016/j.jvc.2012.01.006

[13] Hyun C: Acquired heart valvular diseases. In: Hyun C, editor. Case studies in small animal cardiology, 1st ed. Seoul: Panmun Education; 2013. p. 353–398.

[14] Yosefy C, Ben BA: Floppy mitral valve/mitral valve prolapse and genetics. J Heart Valve Dis. 2007; **16**: 590–595.

[15] Lewis T, Swift S, Woolliams JA, Blott S: Heritability of premature mitral valve disease in cavalier king charles spaniels. Vet J. 2011; **188**: 73–76. doi:10.1016/j.tvjl.2010.02.016

[16] Kienle R, Thomas W. Echocardiography. In: Nyland T, Mattoon J, editors. Small animal diagnostic ultrasound, 2nd ed. Philadelphia: WB Saunders; 2002. p. 354–423.

[17] Das K, Tashjian R: Chronic mitral valve disease in the dog. Vet Med Small Anim Clin. 1965; **60**: 1209–1216.

[18] Kvart C, Häggström J, Pedersen HD, Hansson K, Eriksson A, Järvinen A-K, Tidholm A, Bsenko K, Ahlgren E, Lives M, Åblad B, Falk T, Bjerkås E, Gundler S, Lord P, Wegeland G, Adolfsson E, Corfitzen J: Efficacy of enalapril for prevention of congestive heart failure in dogs with myxomatous valve disease and asymptomatic mitral regurgitation. J Vet Intern Med. 2002; **16**: 80–88. doi:10.1111/j.1939-1676.2002.tb01610.x

[19] Kogure K: Pathology of chronic mitral valvular disease in the dog. Nihon Juigaku Zasshi. 1980; **42**: 323–335. doi:10.1292/jvms1939.42.323

[20] Pomerance A, Whitney JC: Heart valve changes common to man and dog: a comparative study. Cardiovasc Res. 1970; **4**: 61–66. doi:10.1111/j.1748-5827.1970.tb05603.x

[21] Davies M, Moore B, Braimbridge M: The floppy mitral valve. Study of incidence, pathology, and complications in surgical, necropsy, and forensic material. Br Heart J. 1978; **40**: 468. doi:10.1136/hrt.40.5.468

[22] Kihara Y, Sasayama S, Miyazaki S, Onodera T, Susawa T, Nakamura Y, Fujiwara H, Kawai C: Role of the left atrium in adaptation of the heart to chronic mitral regurgitation in conscious dogs. Circ Res. 1988; **62**: 543–553. doi:10.1161/01.RES.62.3.543

[23] Olsen LH, Häggström J, Petersen HD. Acquired valvular heart disease. In: Ettinger SJ, Feldman E, editors. Textbook of veterinary internal medicine, 7th ed. Philadelphia: W.B. Saunders; 2010. p. 1299–1319.

[24] Häggström J, Kvart C, Hansson K: Heart sounds and murmurs: changes related to severity of chronic valvular disease in the cavalier king charles spaniel. J Vet Intern Med. 1995; **9**: 75–85. doi:10.1111/j.1939-1676.1995.tb03276.x

[25] Smith P: Management of chronic degenerative mitral valve disease in dogs. In Practice. 2006; 376. doi:10.1136/inpract.28.7.376

[26] Choi W-J: Evaluation of hepatic function panel in dogs with degenerative valvular disease [thesis]. Chuncheon: Kangwon National University; 2016.

[27] Han D, Choi R, Hyun C: Canine pancreatic-specific lipase concentrations in dogs with heart failure and chronic mitral valvular insufficiency. J Vet Intern Med. 2015; **29**: 180–183. doi:10.1111/jvim.12521

[28] Choi B-S, Moon H-S, Suh S-I, Hyun C. Evaluation of serum cystatin-C and symmetric dimethylarginine concentrations in dogs with heart failure from chronic mitral valvular insufficiency. J Vet Med Sci. 2016, in press.

[29] Maisel AS, Krishnaswamy P, Nowak RM, McCord J, Hollander JE, Duc P, Omland T, Storrow AB, Abraham WT, Wu AH, Clopton P, Steg PG, Westheim A, Knudsen CW, Perez A, Kazanegra R, Herrmann HC, McCullough PA: Rapid measurement of b-type natriuretic peptide in the emergency diagnosis of heart failure. N Engl J Med. 2002; **347**: 161–167. doi:10.1056/NEJMoa020233

[30] Antman EM, Tanasijevic MJ, Thompson B, Schactman M, McCabe CH, Cannon CP, Fischer GA, Fung AY, Thompson C, Wybenga D, Braunwald E: Cardiac-specific troponin

i levels to predict the risk of mortality in patients with acute coronary syndromes. N Engl J Med. 1996; **335**: 1342–1349. doi:10.1056/NEJM199610313351802

[31] Jain MK, Ridker PM: Anti-inflammatory effects of statins: clinical evidence and basic mechanisms. Nat Rev Drug Discov. 2005; **4**: 977–987. doi:10.1038/nrd1901

[32] Spratt DP, Mellanby RJ, Drury N, Archer J: Cardiac troponin I: evaluation i of a biomarker for the diagnosis of heart disease in the dog. J Small Anim Pract. 2005; **46**: 139–145. doi:10.1111/j.1748-5827.2005.tb00304.x

[33] Herndon WE, Kittleson MD, Sanderson K, Drobatz KJ, Clifford CA, Gelzer A, Summerfield NJ, Linde A, Sleeper MM: Cardiac troponin i in feline hypertrophic cardiomyopathy. J Vet Intern Med. 2002; **16**: 558–564. doi:10.1111/j.1939-1676.2002.tb02387.x

[34] Hagman R, Lagerstedt AS, Fransson BA, Bergstrom A, Haggstrom J: Cardiac troponin i levels in canine pyometra. Acta Vet Scand. 2007; **49**: 6. doi:10.1186/1751-0147-49-6

[35] Schober KE, Cornand C, Kirbach B, Aupperle H, Oechtering G: Serum cardiac troponin i and cardiac troponin t concentrations in dogs with gastric dilatation-volvulus. J Am Vet Med Assoc. 2002; **221**: 381–388. doi:10.2460/javma.2002.221.381

[36] Schober KE, Kirbach B, Oechtering G: Noninvasive assessment of myocardial cell injury in dogs with suspected cardiac contusion. J Vet Cardiol. 1999; **1**: 17–25. doi:10.1016/S1760-2734(06)70030-3

[37] Palazzuoli A, Beltrami M, Ruocco G, Pellegrini M, Nuti R: The role of natriuretic peptides for the diagnosis of left ventricular dysfunction. Sci. World J. 2013; 2013: 784670. doi:10.1155/2013/784670

[38] DeFrancesco TC, Rush JE, Rozanski EA, Hansen BD, Keene BW, Moore DT, Atkins CE: Prospective clinical evaluation of an elisa b-type natriuretic peptide assay in the diagnosis of congestive heart failure in dogs presenting with cough or dyspnea. J Vet Intern Med. 2007; **21**: 243–250. doi:10.1111/j.1939-1676.2007.tb02956.x

[39] Boswood A, Dukes-McEwan J, Loureiro J, James RA, Martin M, Stafford-Johnson M, Smith P, Little C, Attree S: The diagnostic accuracy of different natriuretic peptides in the investigation of canine cardiac disease. J Small Anim Pract. 2008; **49**: 26–32. doi:10.1111/j.1748-5827.2007.00510.x

[40] Caspi D, Snel FW, Batt RM, Bennett D, Rutteman GR, Hartman EG, Baltz ML, Gruys E, Pepys MB: C-reactive protein in dogs. Am J Vet Res. 1987; **48**: 919–921.

[41] Kushner I, Feldmann G: Control of the acute phase response. Demonstration of c-reactive protein synthesis and secretion by hepatocytes during acute inflammation in the rabbit. J Exp Med. 1978; **148**: 466–477. doi:10.1084/jem.148.2.466

[42] Dabrowski R, Kostro K, Lisiecka U, Szczubial M, Krakowski L: Usefulness of c-reactive protein, serum amyloid a component, and haptoglobin determinations in bitches with pyometra for monitoring early post-ovariohysterectomy complications. Theriogenology. 2009; **72**: 471–476. doi:10.1016/j.theriogenology.2009.03.017

[43] Kjelgaard-Hansen M, Jensen AL, Houser GA, Jessen LR, Kristensen AT: Use of serum c-reactive protein as an early marker of inflammatory activity in canine type II immune-mediated polyarthritis: case report. Acta Vet Scand. 2006; **48**: 9. doi:10.1186/1751-0147-48-9

[44] Matijatko V, Mrljak V, Kis I, Kucer N, Forsek J, Zivicnjak T, Romic Z, Simec Z, Ceron JJ: Evidence of an acute phase response in dogs naturally infected with babesia canis. Vet Parasitol. 2007; **144**: 242–250. doi:10.1016/j.vetpar.2006.10.004

[45] Nakamura M, Takahashi M, Ohno K, Koshino A, Nakashima K, Setoguchi A, Fujino Y, Tsujimoto H: C-reactive protein concentration in dogs with various diseases. J Vet Med Sci. 2008; **70**: 127–131. doi:10.1292/jvms.70.127

[46] Otabe K, Ito T, Sugimoto T, Yamamoto S: C-reactive protein (CRP) measurement in canine serum following experimentally-induced acute gastric mucosal injury. Lab Anim. 2000; **34**: 434–438. doi:10.1258/002367700780387679

[47] Yamamoto S, Shida T, Miyaji S, Santsuka H, Fujise H, Mukawa K, Furukawa E, Nagae T, Naiki M: Changes in serum c-reactive protein levels in dogs with various disorders and surgical traumas. Vet Res Commun. 1993; **17**: 85–93. doi:10.1007/BF01839236

[48] Rush JE, Lee ND, Freeman LM, Brewer B: C-reactive protein concentration in dogs with chronic valvular disease. J Vet Intern Med. 2006; **20**: 635–639. doi:10.1111/j.1939-1676.2006.tb02908.x

[49] Hyun C. Valvular diseases. In: Hyun C, editor. Small animal cardiology, 1st ed. Seoul: Panmun Publication; 2005. p. 185–206.

[50] Häggström J, Hamlin RL, Hansson K, Kvart C: Heart rate variability in relation to severity of mitral regurgitation in cavalier king charles spaniels. J Small Anim Pract. 1996; **37**: 69–75. doi:10.1111/j.1748-5827.1996.tb01941.x

[51] Lopez-Alvarez J, Boswood A, Moonarmart W, Hezzell MJ, Lotter N, Elliott J: Longitudinal electrocardiographic evaluation of dogs with degenerative mitral valve disease. J Vet Intern Med. 2014; **28**: 393–400. doi:10.1111/jvim.12311

[52] Rush JE: Chronic valvular heart disease in dogs. In: Proceedings of 26th Annual Waltham Diets/OSU Symposium for the Treatment of Small Animal Cardiology. 2002.

[53] Buchanan JW, Bucheler J: Vertebral scale system to measure canine heart size in radiographs. J Am Vet Med Assoc. 1995; **206**: 194–199.

[54] Lord PF, Hansson K, Carnabuci C, Kvart C, Haggstrom J: Radiographic heart size and its rate of increase as tests for onset of congestive heart failure in cavalier king charles spaniels with mitral valve regurgitation. J Vet Intern Med. 2011; **25**: 1312–1319. doi:10.1111/j.1939-1676.2011.00792.x

[55] Ettinger S, Kantrowitz B. Disease of the trachea. In: Ettinger S, Feldman E, editors. Textbook of veterinary internal medicine. 6th ed. St. Louis, MO: Elsevier Saunders; 2005. p. 1217–1232.

[56] Boon JA: Acquired heart disease: mitral insufficiency; In: Boon JA, editor. Manual of veterinary echocardiography. 1st ed. Baltimore: Williams and Wilkins; 1998. p. 261–286.

[57] Serres FJ, Chetboul V, Tissier R, Carlos Sampedrano C, Gouni V, Nicolle AP, Pouchelon JL: Doppler echocardiography-derived evidence of pulmonary arterial hypertension in dogs with degenerative mitral valve disease: 86 cases (2001–2005). J Am Vet Med Assoc. 2006; **229**: 1772–1778. doi:10.2460/javma.229.11.1772

[58] Stepien RL: Pulmonary arterial hypertension secondary to chronic left-sided cardiac dysfunction in dogs. J Small Anim Pract. 2009; **50**(Suppl 1): 34–43. doi:10.1111/j.1748-5827.2009.00802.x

[59] Chiavegato D, Borgarelli M, D'Agnolo G, Santilli RA: Pulmonary hypertension in dogs with mitral regurgitation attributable to myxomatous valve disease. Vet Radiol Ultrasound. 2009; **50**: 253–258. doi:10.1111/j.1740-8261.2009.01529.x

[60] Bonagura JD, Schober KE: Can ventricular function be assessed by echocardiography in chronic canine mitral valve disease? J Small Anim Pract. 2009; **50**(Suppl 1): 12–24. doi:10.1111/j.1748-5827.2009.00803.x

[61] Serres F, Chetboul V, Tissier R, Poujol L, Gouni V, Carlos Sampedrano C, Pouchelon JL: Comparison of 3 ultrasound methods for quantifying left ventricular systolic function: correlation with disease severity and prognostic value in dogs with mitral valve disease. J Vet Intern Med. 2008; **22**: 566–577. doi:10.1111/j.1939-1676.2008.0097.x

[62] Kvart C, Häggström J. Acquired valvular heart disease. In: Ettinger S, Feldman E, editors. Textbook of veterinary internal medicine, 6th ed. Philadelphia: WB Saunders; 2005. p. 1022–1039.

[63] Bonagura JD, Herring DS: Echocardiography. Acquired heart disease. Vet Clin North Am Small Anim Pract. 1985; **15**: 1209–1224. doi:10.1016/S0195-5616(85)50366-6

[64] Terzo E, Di Marcello M, McAllister H, Glazier B, Lo Coco D, Locatelli C, Palermo V, Brambilla PG: Echocardiographic assessment of 537 dogs with mitral valve prolapse and leaflet involvement. Vet Radiol Ultrasound. 2009; **50**: 416–422. doi:10.1111/j.1740-8261.2009.01559.x

[65] Olivier N, Kittleson M, Eyster G, Miller J: M-mode echocardiography in the diagnosis of ruptured mitral chordae tendineae in a dog. J Am Vet Med Assoc. 1984; **184**: 588–589.

[66] Jacobs G, Calvert C, Mahaffey M, Hall D: Echocardiographic detection of flail left atrioventricular valve cusp from ruptured chordae tendineae in 4 dogs. J Vet Intern Med. 1995; **9**: 341–346. doi:10.1111/j.1939-1676.1995.tb01095.x

[67] Kittleson M, Kienle R. Myxomatous atrioventricular valvular degeneration. In: Kittleson M, Kienle R, eds. Small animal cardiovascular medicine. St. Louis: Mosby; 1998. p. 297–318.

[68] Serres F, Chetboul V, Tissier R, Sampedrano CC, Gouni V, Nicolle AP, Pouchelon JL: Chordae tendineae rupture in dogs with degenerative mitral valve disease: prevalence,

survival, and prognostic factors (114 cases, 2001–2006). J Vet Intern Med. 2007; **21**: 258–264. doi:10.1111/j.1939-1676.2007.tb02958.x

[69] Muzzi RA, de Araujo RB, Muzzi LA, Pena JL, Silva EF: Regurgitant jet area by doppler color flow mapping: quantitative assessment of mitral regurgitation severity in dogs. J Vet Cardiol. 2003; **5**: 33–38. doi:10.1016/S1760-2734(06)70050-9

[70] Gouni V, Serres FJ, Pouchelon JL, Tissier R, Lefebvre HP, Nicolle AP, Sampedrano CC, Chetboul V: Quantification of mitral valve regurgitation in dogs with degenerative mitral valve disease by use of the proximal isovelocity surface area method. J Am Vet Med Assoc. 2007; **231**: 399–406. doi:10.2460/javma.231.3.399

[71] Zoghbi WA, Enriquez-Sarano M, Foster E, Grayburn PA, Kraft CD, Levine RA, Nihoyannopoulos P, Otto CM, Quinones MA, Rakowski H, Stewart WJ, Waggoner A, Weissman NJ: Recommendations for evaluation of the severity of native valvular regurgitation with two-dimensional and doppler echocardiography. J Am Soc Echocardiogr. 2003; **16**: 777–802. doi:10.1016/S0894-7317(03)00335-3

[72] Patel AR, Mochizuki Y, Yao J, Pandian NG: Mitral regurgitation: comprehensive assessment by echocardiography. Echocardiography. 2000; **17**: 275–283. doi:10.1111/j.1540-8175.2000.tb01138.x

[73] Triboulloy C, Shen W, Quere J, Rey J, Choquet D, Dufossé H, Lesbre J: Assessment of severity of mitral regurgitation by measuring regurgitant jet width at its origin with transesophageal Doppler color flow imaging. Circulation. 1992; **85**: 1248–1253. doi:10.1161/01.CIR.85.4.1248

[74] Fehske W, Omran H, Manz M, Köhler J, Hagendorff A, Lüderitz B: Color-coded Doppler: imaging of the vena contracta as a basis of quantification of pure mitral regurgitation. Am J Cardiol. 1994; **73**: 268–274. doi:10.1016/0002-9149(94)90232-1

[75] Little SH, Igo SR, Pirat B, McCulloch M, Hartley CJ, Nosé Y, Zoghbi WA: In vitro validation of real-time three-dimensional color Doppler echocardiography for direct measurement of proximal isovelocity surface area in mitral regurgitation. Am J Cardiol. 2007; **99**: 1440–1447. doi:10.1016/j.amjcard.2006.12.079

[76] Kittleson MD, Brown WA: Regurgitant fraction measured by using the proximal isovelocity surface area method in dogs with chronic myxomatous mitral valve disease. J Vet Intern Med. 2003; **17**: 84–88. doi:10.1892/0891-6640(2003)017<0084:RFMBUT>2.3.CO;2

[77] Peddle GD, Buchanan JW: Acquired atrial septal defects secondary to rupture of the atrial septum in dogs with degenerative mitral valve disease. J Vet Cardiol. 2010; **12**: 129–134. doi:10.1016/j.jvc.2010.03.002

[78] Buchanan JW, Kelly AM: Endocardial splitting of the left atrium in the dog with hemorrhage and hemopericardium. J AM Vet Rad Soc. 1964; **5**: 28–39. doi:10.1111/j.1740-8261.1964.tb01302.x

[79] Sadanaga KK, MacDonald MJ, Buchanan JW: Echocardiography and surgery in a dog with left atrial rupture and hemopericardium. J Vet Intern Med. 1990; **4**: 216–221. doi:10.1111/j.1939-1676.1990.tb00900.x

[80] Katayama K, Tajimi T, Guth BD, Matsuzaki M, Lee JD, Seitelberger R, Peterson KL: Early diastolic filling dynamics during experimental mitral regurgitation in the conscious dog. Circulation. 1988; **78**: 390–400. doi:10.1161/01.CIR.78.2.390

[81] O'Gara P, Sugeng L, Lang R, Sarano M, Hung J, Raman S, Fischer G, Carabello B, Adams D, Vannan M: The role of imaging in chronic degenerative mitral regurgitation. JACC Cardiovasc Imaging. 2008; **1**: 221–237. doi:10.1016/j.jcmg.2008.01.011

[82] Richard D, William P. Thomas. Echocardiography. In: Nyland TG, Mattoon JS, editors. Small animal diagnostic ultrasound, 2nd ed. Philadelphia: WB Saunders; 2002. p. 370.

[83] Kittleson MD, Eyster GE, Knowlen GG, Bari Olivier N, Anderson LK: Myocardial function in small dogs with chronic mitral regurgitation and severe congestive heart failure. J Am Vet Med Assoc. 1984; **184**: 455–459.

[84] Chetboul V, Tissier R: Echocardiographic assessment of canine degenerative mitral valve disease. J Vet Cardiol. 2012; **14**: 127–148. doi:10.1016/j.jvc.2011.11.005

[85] Chetboul V, Carlos Sampedrano C, Gouni V, Nicolle AP, Pouchelon JL, Tissier R: Ultrasonographic assessment of regional radial and longitudinal systolic function in healthy awake dogs. J Vet Intern Med. 2006; **20**: 885–893. doi:10.1111/j.1939-1676.2006.tb01802.x

[86] Chetboul V: Advanced techniques in echocardiography in small animals. Vet Clin North Am Small Anim Pract. 2010; **40**: 529–543. doi:10.1016/j.cvsm.2010.03.007

[87] Artis NJ, Oxborough DL, Williams G, Pepper CB, Tan LB: Two-dimensional strain imaging: a new echocardiographic advance with research and clinic applications. Int J Cardiol. 2008; **123**: 240–248. doi:10.1016/j.ijcard.2007.02.046

[88] Nagueh SF, Middleton KJ, Kopelen HA, Zoghbi WA, Quiñones MA: Doppler tissue imaging: a noninvasive technique for evaluation of left ventricular relaxation and estimation of filling pressures. J Am Coll Cardiol. 1997; 30: 1527–1533. doi:10.1016/S0735-1097(97)00344-6

[89] Ommen SR, Nishimura RA, Appleton CP, Miller FA, Oh JK, Redfield MM, Tajik AJ: Clinical utility of Doppler echocardiography and tissue Doppler imaging in the estimation of left ventricular filling pressures: a comparative simultaneous Doppler-catheterization study. Circulation. 2000; **102**: 1788–1794. doi:10.1161/01.CIR.102.15.1788

[90] Federman M, Hess O: Differentiation between systolic and diastolic dysfunction. Eur Heart J. 1994; **15**(Suppl D): 2–6. doi:10.1093/eurheartj/15.suppl_D.2

[91] DeMaria A, Wisenbaugh T, Smith M, Harrison M, Berk M: Doppler echocardiographic evaluation of diastolic function. Circulation. 1991; **84**(Suppl 1): 288–295.

[92] Abott JA. Acquired valvular disease. In: Larry P, Tilley FWKSJ, Oyama MA, Meg MS, editors. Manual of canine and feline cardiology, 4th ed. Philadelphia: WB Saunders; 2008. p. 110–138.

[93] Strickland KN. Pathophysiology and therapy of heart failure. In: Larry P, Tilley FWKSJ, Oyama MA, Meg MS, editors. Manual of canine and feline cardiology, 4th ed. Philadelphia: WB Saunders; 2008. p. 288–314.

[94] Schober KE, Bonagura JD, Scansen BA, Stern JA, Ponzio NM: Estimation of left ventricular filling pressure by use of doppler echocardiography in healthy anesthetized dogs subjected to acute volume loading. Am J Vet Res. 2008; **69**: 1034–1049. doi:10.2460/javma.233.4.617

[95] Schober KE, Stern JA, DaCunha DN, Pedraza-Toscano AM, Shemanski D, Hamlin RL: Estimation of left ventricular filling pressure by doppler echocardiography in dogs with pacing-induced heart failure. J Vet Intern Med. 2008; **22**: 578–585. doi:10.1111/j.1939-1676.2008.0099.x

[96] Ho D-M. Evaluation of prognostic echocardiographic markers in Dogs with Chronic Mitral Valve Insufficiency [thesis]. Chuncheon: Kangwon National University; 2016.

[97] BENCH Study Group: The effect of benazepril on survival times and clinical signs of dogs with congestive heart failure: results of a multicenter, prospective, randomized, double-blinded, placebo-controlled, long-term clinical trial. J Vet Cardiol. 1999; **1**: 7–18. doi:10.1016/S1760-2734(06)70025-X

[98] Ettinger SJ, Benitz AM, Ericsson GF, Cifelli S, Jernigan AD, Longhofer SL, Trimboli W, Hanson PD: Effects of enalapril maleate on survival of dogs with naturally acquired heart failure. The long-term investigation of veterinary enalapril (live) study group. J Am Vet Med Assoc. 1998; **213**: 1573–1577.

[99] Häggström J, Boswood A, O'Grady M, Jöns O, Smith S, Swift S, Borgarelli M, Gavaghan B, Kresken JG, Patteson M, Ablad B, Bussadori CM, Glaus T, Kovacević A, Rapp M, Santilli RA, Tidholm A, Eriksson A, Belanger MC, Deinert M, Little CJ, Kvart C, French A, Rønn-Landbo M, Wess G, Eggertsdottir AV, O'Sullivan ML, Schneider M, Lombard CW, Dukes-McEwan J, Willis R, Louvet A, DiFruscia R: Effect of pimobendan or benazepril hydrochloride on survival times in dogs with congestive heart failure caused by naturally occurring myxomatous mitral valve disease: the QUEST study. J Vet Intern Med. 2008; **22**: 1124–1135. doi:10.1111/j.1939-1676.2008.0150.x

[100] Lombard CW, Jons O, Bussadori CM: Clinical efficacy of pimobendan versus benazepril for the treatment of acquired atrioventricular valvular disease in dogs. J Am Anim Hosp Assoc. 2006; **42**: 249–261. doi:10.5326/0420249

[101] Woodfield JA: Controlled clinical evaluation of enalapril in dogs with heart failure: results of the cooperative veterinary enalapril study group the cove study group. J Vet Intern Med. 1995; **9**: 243–252. doi:10.1111/j.1939-1676.1995.tb01075.x

[102] Avalonmed. Mitralseal canine mitral valve replacement technology. 2011. Available from: http://www.youtube.com/watch?v=SbEBsejEJ1I [Accessed: 2016-08-24].

[103] Scientific Animations: Mitral valve repair surgery—cardiac animation. 2011. Available from: http://www.youtube.com/watch?v=t9FpQ_wSVD4 [Accessed: 2016-08-24].

[104] Atkins CE, Rausch WP, Gardner SY, Defrancesco TC, Keene BW, Levine JF: The effect of amlodipine and the combination of amlodipine and enalapril on the renin-angiotensin-aldosterone system in the dog. J Vet Pharmacol Ther. 2007; **30**: 394–400. doi:10.1111/j.1365-2885.2007.00894.x

[105] Vollmar AC, Fox PR: Long-term outcome of Irish Wolfhound Dogs with preclinical cardiomyopathy, atrial fibrillation, or both treated with Pimobendan, Benazepril Hydrochloride, or Methyldigoxin Monotherapy. J Vet Intern Med. 2016; **30**: 553–559. doi:10.1111/jvim.13914

[106] Park D-S. Evaluation of vital signs and echocardiographic indices in asymptomatic dogs with chronic mitral valvular insufficiency after efficacy of pimobendan and enalapril treatment [thesis]. Chuncheon: Kangwon National University; 2016.

[107] Ouellet M, Bélanger MC, Difruscia R, Beauchamp G: Effect of pimobendan on echocardiographic values in dogs with asymptomatic mitral valve disease. J Vet Intern Med. 2009; **23**: 258–263. doi:10.1111/j.1939-1676.2008.0239.x

[108] Thadani U, Fung HL, Darke AC, Parker JO: Oral isosorbide dinitrate in the treatment of angina pectoris. Dose-response relationship and duration of action during acute therapy. Circulation. 1980; **62**: 491–502. doi:10.1161/01.CIR.62.3.491

[109] Schroeder JS: Combination therapy with isosorbide dinitrate: current status and the future. Am Heart J. 1985; **110**: 284–291. doi:10.1016/0002-8703(85)90503-4

[110] Aviado DM, Folle LE, Bellet S: Cardiopulmonary effects of glyceryl trinitrate and isosorbide dinitrate. Cardiologia. 1968; **52**: 287–303. doi:10.1159/000166129

[111] Cohn JN, Johnson G, Ziesche S, Cobb F, Francis G, Tristani F, Smith R, Dunkman WB, Loeb H, Wong M, Bhat G, Goldman S, Fletcher RD, Doherty J, Hughes CV, Carson P, Cintron G, Shabetai R, Haakenson C: A comparison of enalapril with hydralazine-isosorbide dinitrate in the treatment of chronic congestive heart failure. N Engl J Med. 1991; **325**: 303–310. doi:10.1056/NEJM199108013250502

[112] McDonald KM, Francis GS, Matthews J, Hunter D, Cohn JN: Long-term oral nitrate therapy prevents chronic ventricular remodeling in the dog. J Am Coll Cardiol. 1993; 21: 514–522. doi:10.1016/0735-1097(93)90697-Y

[113] Kim J-B, Suh S-I, Hyun C: Evaluation of efficacy of sustained-release form of isosorbide dinitrate in dogs heart failure from chronic mitral valvular insufficiency. J Vet Clin. 2016; in press.

[114] Tashjian RJ, Das KM, Palich WE, Hamlin RL, Yarns DA: Studies on cardiovascular disease in the cat. Ann NY Acad Sci. 1965; **127**: 581–605. doi:10.1111/j.1749-6632.1965.tb49425.x

[115] Thomason JD, Fallaw TK, Calvert C: Pimobendan treatment in dogs with congestive heart failure. Vet Med. 2007; **102**: 736–740.

[116] Gordon SG, Saunders AB, Roland RM, Winter RL, Drourr L, Achen SE, Hariu CD, Fries RC, Boggess MM, Miller MW: Effect of oral administration of pimobendan in cats with heart failure. J Am Vet Med Assoc. 2012; **241**: 89–94. doi:10.2460/javma.241.1.89

[117] Ishiki R, Ishihara T, Izawa H, Nagata K, Hirai M, Yokota M: Acute effects of a single low oral dose of pimobendan on left ventricular systolic and diastolic function in patients with congestive heart failure. J Cardiovasc Pharmacol. 2000; **35**: 897–905. doi:10.1097/00005344-200006000-00011

[118] Bowles D, Fry D: Pimobendan and its use in treating canine congestive heart failure. Compend Contin Educ Vet. 2011; **33**: E1.

[119] Macgregor JM, Rush JE, Laste NJ, Malakoff RL, Cunningham SM, Aronow N, Hall DJ, Williams J, Price LL: Use of pimobendan in 170 cats (2006–2010). J Vet Cardiol. 2011; **13**: 251–260. doi:10.1016/j.jvc.2011.08.001

[120] Asanoi H, Ishizaka S, Kameyama T, Ishise H, Sasayama S: Disparate inotropic and lusitropic responses to pimobendan in conscious dogs with tachycardia-induced heart failure. J Cardiovasc Pharmacol. 1994; **23**: 268–274. doi:10.1097/00005344-199411000-00014

[121] Fuentes VL, Corcoran B, French A, Schober KE, Kleemann R, Justus C: A double-blind, randomized, placebo-controlled study of pimobendan in dogs with dilated cardiomyopathy. J Vet Intern Med. 2002; **16**: 255–261. doi:10.1111/j.1939-1676.2002.tb02366.x

[122] Hanzlicek AS, Gehring R, Kukanich B, Kukanich KS, Borgarelli M, Smee N, Olson EE, Margiocco M: Pharmacokinetics of oral pimobendan in healthy cats. J Vet Cardiol. 2012; **14**: 489–496. doi:10.1016/j.jvc.2012.06.002

[123] Fusellier M, Desfontis JC, Le Roux A, Madec S, Gautier F, Thuleau A, Gogny M: Effect of short-term treatment with meloxicam and pimobendan on the renal function in healthy beagle dogs. J Vet Pharmacol Ther. 2008; **31**: 150–155. doi:10.1111/j.1365-2885.2007.00934.x

[124] Reina-Doreste Y, Stern JA, Keene BW, Tou SP, Atkins CE, DeFrancesco TC, Ames MK, Hodge TE, Meurs KM: Case-control study of the effects of pimobendan on survival time in cats with hypertrophic cardiomyopathy and congestive heart failure. J Am Vet Med Assoc. 2014; **245**: 534–539. doi:10.2460/javma.245.5.534

[125] Hambrook LE, Bennett PF: Effect of pimobendan on the clinical outcome and survival of cats with non-taurine responsive dilated cardiomyopathy. J Feline Med Surg. 2012; **14**: 233–239. doi:10.1177/1098612X11429645

[126] Kanno N, Kuse H, Kawasaki M, Hara A, Kano R, Sasaki Y: Effects of pimobendan for mitral valve regurgitation in dogs. J Vet Med Sci. 2007; **69**: 373–377. doi:10.1292/jvms.69.373

[127] Reinker LN, Lee JA, Hovda LR, Rishniw M: Clinical signs of cardiovascular effects secondary to suspected pimobendan toxicosis in five dogs. J Am Anim Hosp Assoc. 2012; **48**: 250–255. doi:10.5326/JAAHA-MS-5775

[128] Tissier R, Chetboul V, Moraillon R, Nicolle A, Carlos C, Enriquez B, Pouchelon JL: Increased mitral valve regurgitation and myocardial hypertrophy in two dogs with long-term pimobendan therapy. Cardiovasc Toxicol. 2005; **5**: 43–51. doi:10.1385/CT:5:1:043

[129] Nicolle AP, Chetboul V, Allerheiligen T, Pouchelon JL, Gouni V, Tessier-Vetzel D, Sampedrano CC, Lefebvre HP: Azotemia and glomerular filtration rate in dogs with chronic valvular disease. J Vet Intern Med. 2007; **21**: 943–949. doi:10.1111/j.1939-1676.2007.tb03047.x

[130] Braunwald E: Valvular heart disease. In: Braunwald E, editor. Heart disease, 5th ed. Philadelphia: Saunders; 1997. p. 1017.

[131] Kellum HB, Stepien RL: Sildenafil citrate therapy in 22 dogs with pulmonary hypertension. J Vet Intern Med. 2007; **21**: 1258–1264. doi:10.1892/07-006.1

[132] Moreno HJ: Clinical characteristics and outcome of 13 dogs with pulmonary arterial hypertension (ph) treated with sildenafil. J Vet Intern Med. 2007; **21**: 1165. doi:10.1892/0891-6640(2007)21[1165:C]2.0.CO;2

[133] Sisson D, Kvart C, Darke P. Acquired valvular heart disease in dogs and cats. In: Fox P, Sisson D, Moise N, editors. Textbook of canine and feline cardiology, 1st ed. Philadelphia: Saunders; 1999. p. 536.

[134] Bock JS, Gottlieb SS: Cardiorenal syndrome: new perspectives. Circulation. 2010; **121**: 2592–2600. doi:10.1161/CIRCULATIONAHA.109.886473

[135] Watson P: Chronic pancreatitis in dogs. Top Companion Anim Med. 2012; **27**: 133–139. doi:10.1053/j.tcam.2012.04.006

[136] Scopinaro N, Gianetta E, Civalleri D, Bonalumi U, Bachi V: Bilio-pancreatic bypass for obesity: 1. An experimental study in dogs. Br J Surg. 1979; **66**: 613–617. doi:10.1002/bjs.1800660905

[137] Webster CRL: History, clinical signs, and physical findings in hepatobiliary disease. In Ettinger S, Feldman E (eds): Textbook of veterinary internal medicine, 7th ed. Philadelphia: W.B. Saunders; 2010: p. 1612–1625.

[138] Steiner JM. Canine pancreatic disease. In: Ettinger SJ, Feldman EC, editors. Textbook of veterinary internal medicine, 7th ed. Philadelphia: W.B. Saunders; 2010. p. 1695–1703.

[139] Borgarelli M, Häggström J: Canine degenerative myxomatous mitral valve disease: natural history, clinical presentation and therapy. Vet Clin North Am Small Anim Pract. 2010; **40**: 651–663. doi:10.1016/j.cvsm.2010.03.008

[140] Gullo L, Cavicchi L, Tomassetti P, Spagnolo C, Freyrie A, D'Addato M: Effects of ischemia on the human pancreas. Gastroenterology. 1996; **111**: 1033–1038. doi:10.1016/S0016-5085(96)70072-0

[141] Hackert T, Hartwig W, Schneider L, Strobel O, Büchler M, Werner J: Ischemic acute pancreatitis: clinical features, diagnosis, therapy and outcome. Pancreas. 2007; **35**: 406. doi:10.1097/01.mpa.0000297712.13843.c3

[142] Moneva-Jordan A: Dietary considerations for the cardiac patient. In Practice. 2012; **38**: 28–29. doi:10.1136/inp.f6018

[143] Sakorafas GH, Tsiotos GG, Sarr MG: Ischemia/reperfusion-induced pancreatitis. Dig Surg. 2000; **17**: 3–14. doi:10.1159/000018793

[144] Warshaw AL, O'Hara PJ: Susceptibility of the pancreas to ischemic injury in shock. Ann Surg. 1978; **188**: 197–201. doi:10.1097/00000658-197808000-00012

[145] Seo JI, Han SH, Choi R, Han J, Lee L, Hyun C: Cardiopulmonary and anesthetic effects of the combination of butorphanol, midazolam and alfaxalone in Beagle dogs. Vet Anaesth Analg. 2015; **42**: 304–308. doi:10.1111/vaa.12223

[146] Fox PR, Sisson D, Moise NS. International small animal cardiac health council. Appendix A—recommendations for diagnosis of heart disease and treatment of heart failure in small animals. In: Fox PR, Sisson D, Moise NS, editors. Textbook of canine and feline cardiology, 2nd ed. Philadelphia: W.B. Saunders; 1999. p. 883–901.

[147] Pape LA, Price JM, Alpert JS, Ockene IS, Weiner BH: Relation of left atrial size to pulmonary capillary wedge pressure in severe mitral regurgitation. Cardiology. 1991; **78**: 297–303. doi:10.1159/000174808

[148] Hezzell MJ, Boswood A, Moonarmart W, Elliott J: Selected echocardiographic variables change more rapidly in dogs that die from myxomatous mitral valve disease. J Vet Cardiol. 2012; **14**: 269–279. doi:10.1016/j.jvc.2012.01.009

Section 4

Reproduction

Chapter 6

Abnormalities in the Sexual Cycle of Bitches

Ali Risvanli, Halis Ocal and Cahit Kalkan

Additional information is available at the end of the chapter

http://dx.doi.org/10.5772/64648

Abstract

Sexual-cycle abnormalities are an important cause of infertility in bitches, with disorders such as anestrus, split estrus, and persistent estrus having varied etiologies. Sexual-cycle abnormalities in bitches may be addressed as follicular- or luteal-phase disorders. However, pet owners should have a good working knowledge of the sexual cycles of their animals in order to better understand these disorders.

Keywords: abnormality, sexual cycle, bitch

1. Introduction

Sexual-cycle abnormalities in bitches may present as anestrous, shorter, or longer cycles, as well as prolonged proestrus, prolonged estrus, split estrus, or anovulatory cycles. These cycle disorders may result from abnormal ovarian functions and are a cause of infertility [1].

1.1. Anestrous

Anestrous cycles in bitches may be either primary or secondary. If a bitch does not show signs of estrus despite having reached age of puberty, anestrous cycles are primary. The age of puberty in bitches is 6–14 months. In general, the "primary anestrus" diagnosis may be used if estrus has not occurred and the cycle has not started by 24 months of age. Although some small breeds experience first estrus at 6 months of age, cycles accompanied by estrus signs may be delayed, since the first cycles may be silent. Therefore, cycle problems are not usually investigated until a bitch reaches 2 years of age. A diagnosis of "secondary estrus" is used if estrus has not occurred for 10–18 months, although first estrus had occurred. In other words, secondary estrus is defined as the presence of a period longer than 10–18 months between estruses. Normally, bitches experience estrus at 4–10 month (mean 6–7 months) intervals. Cycles shorter than 4 months or longer than 10 months are abnormal and can cause infertility [2–6].

Factors that lead to primary anestrus include ovariectomy or ovariohysterectomy at early age, silent heat (subestrus), abnormalities in sexual differentiation (chromosomal and genetic disorders), use of progesterone or glucocorticoids, congenital hypothyroidism, certain systemic diseases, ovarian anomalies or ovarian aplasia, progesterone-releasing ovarian cysts, and autoimmune oophoritis [2, 7].

Secondary anestrus may result from dysfunction of the thyroid gland or adrenal cortex, as well as nonendocrinological disorders, cachexia, obesity, and use of cycle-inhibiting drugs. Silent heat, luteal cysts, and some ovarian tumors may also lead to secondary anestrus as well as to primary anestrus [4, 5].

1.1.1. History of ovariohysterectomy

If taken from another person, the bitch may have been sterilized, which a new owner may overlook. Presence of a tattoo in the inguinal region and palpation or inspection of a scar from operation in the ventral wall of the abdomen may be indicators of ovariohysterectomy. However, it should be kept in mind that such a scar may be present if any intra-abdominal operation has been performed, so it may be premature to conclude that the scar resulted from ovariohysterectomy. Serum LH level measurement may be used to identify bitches that have undergone ovariohysterectomy, as serum LH is continuously high in such bitches as a result of the absence of negative feedback on LH (because the ovaries have been removed). Although elevation in serum LH provides information about ovariohysterectomy, note that this indicator may also be seen in ovarian dysfunction or during the preovulatory period (i.e., the preovulatory LH peak). Therefore, repeated measurements are required to confirm and experimental laparotomy may also be performed for a definitive diagnosis [8]. Whether ovariohysterectomy has been performed may also be detected by measuring the serum estrogen level before and 60–90 min after intravenous administration of 0.02–0.03 μg/kg buserelin. If ovariohysterectomy has been performed, estrogen levels will be found above 15–20 pg/mL.

Anti-müllerian hormone (AMH) measurement may also show whether a bitch has undergone ovariohysterectomy. AMH levels are found to be significantly lower in bitches that have undergone ovariohysterectomy compared to nonsterilized bitches [9, 10].

1.1.2. Silent heat (subestrus)

Silent heat is defined as the maintenance of ovarian functions without the presence of vulvar edema, serosanguinous vaginal discharge, and charm for male dogs. Silent heat may be observed for several cycles before first estrus in younger bitches of smaller breeds. These animals may be evaluated as "anestrus" because the pet owner may not find the external signs of estrus, or they may be identified as healthy male dogs although ovarian functions continue normally.

If silent heat is suspected, serum progesterone level should be measured once monthly in order to verify that the ovaries are functioning. A serum progesterone level above 2 ng/mL indicates functional luteal tissue. Observation of increasing superficial epithelial cells in regular vaginal cytological examinations is an indicator of functional ovaries [5].

1.1.3. *Disorders of sexual development*

Normal sexual development occurs in three stages: (1) chromosomal (genetic) sex development, (2) gonadal sex development, and (3) phenotypic sex development. Therefore, disorders of sex development and differentiation are classified into three matching groups: (1) sex chromosome disorders, (2) gonadal sex development disorders, and (3) phenotypic sex development disorders. All three groups of disorders result in abnormal sex differentiation and may vary in presentation between genital structure of normal and obscure appearance, and all three groups of disorders may lead to sterility or infertility [11–13].

Chromosomal analysis, anatomical and histopathological definitions of the gonads, and examination of the internal and external genitalia are required for the definite diagnosis of dogs suspected to have disorders of sexual development [11, 14].

1.1.4. *Drug-related anestrus*

Long-term use of some drugs, such as androgen and progestogens, causes anestrus by inhibiting cycles. Exogenous glucocorticoid administration is also reported to affect serum LH level and the normal cycle. Puberty is inhibited due to suppression of genital-canal development and ovarian activity in bitches that have undergone administration of long-acting GnRH agonists like deslorelin [15].

Similarly, cyclic activity may not be observed for a long time in prepubertal bitches that have been actively immunized against GnRH [16].

A comprehensive anamnesis should be obtained from the pet owner if drug-related anestrus is suspected; sufficient data must be obtained regarding the medical history of the bitch. If a pet owner has recently acquired the bitch and if he or she has no or insufficient information about previous vaccinations and medications, drug-related anestrus should always be considered. In such situations, the only treatment is to wait until the effects of the drug(s) disappear or until the antibody titration decreases, if immunized against GnRH.

1.1.5. *Thyroid dysfunctions*

There is an indirect and strict association between thyroid dysfunction and reproductivity. Hypothyroidism leads to reproductive disorders such as prolonged anestrus, silent heat, prolonged proestrus, and ovulatory problems. The prolactin level increases, which leads to impair or no development of ovarian follicles by inhibiting GnRH in bitches that have insufficient thyroid hormone release [17].

In bitches, hypothyroidism usually manifests as primary hypothyroidism, resulting from destruction of the thyroid gland. The serum total and free thyroxin levels are low in bitches with thyroid dysfunction. Thyrotropin-releasing hormone (TRH) secretion from the hypothalamus increases due to low levels of thyroid hormones; consequently, thyroid-stimulating hormone (TSH) secretion from the pituitary gland increases. As a result of this totally physiological process, low serum total and free thyroxin levels and elevated serum TSH levels are observed in bitches with primary hypothyroidism [7].

Hypothyroidism should be kept in mind in the presence of numbness or mental fatigue, hair loss, weight increase or obesity, dryness or loss of body hair, hyperpigmentation, cold intolerance, bradycardia, high plasma cholesterol level, or anemia in a bitch with an anestrous problem. Measurement of only thyroid hormones may yield misleading or conflicting results. Therefore, a full thyroid profile should be obtained and a definitive diagnosis made in bitches suspected of hypothyroidism. For this purpose, the serum-free thyroid hormone level (particularly T4 measurement) and the response of the thyroid gland to TSH administrations should be investigated. Furthermore, autoantibodies against thyroxin (T4) and triiodothyronine (T3) or thyroglobulin should be investigated in serum, as they are produced in most bitches with lymphocytic thyroiditis [7, 18, 19].

The sexual cycle may return to normal with hormone replacement therapy within 3–6 months in bitches diagnosed with hypothyroidism. For this purpose, synthetic thyroid hormone (levothyroxin) should be administered through the peroral route. A dose of 22 μg/kg b.i.d. is usually sufficient.

Although hyperthyroidism with consequent primary anestrus is a rare condition in bitches, a case has been reported in a Pinscher dog with diet-related hyperthyroidism in which primary anestrus developed; the bitch reached proestrus 13 days later after dietary regulation, with estrus then induced by cabergoline [20].

1.1.6. Pituitary gland insufficiency

The pituitary gland is important for the endocrinological functions of the adrenal glands, thyroid gland, and the ovaries. Abnormalities of the pituitary gland also negatively affect these organs. Prolonged anestrus is unusually seen in bitches with dwarfism caused by a congenital anomaly. Ovariohysterectomy is recommended in bitches that show prolonged anestrus related to pituitary gland insufficiency [6].

1.1.7. Systemic diseases

Such diseases may negatively affect reproductive function. Cycles probably will not develop if an animal is unhealthy.

1.1.8. Ovarian anomalies

Progesterone-releasing ovarian cysts, ovarian aplasia, and oophoritis may lead to primary anestrus. Definitive diagnosis can be made by histopathological examination of the ovarian tissue [21].

http://dx.doi.org/10.5772/64648

1.2. Approach to bitches with anestrous problems

First, it should be determined whether the pet owner has correct and sufficient knowledge of the cycle and estrous signs of bitches. The age of the bitch and whether it has experienced ovariectomy, ovariohysterectomy, or administration of any drug or vaccine (especially GnRH vaccine) should be determined. Serum progesterone level should be monitored monthly for 6–8 months before any intervention in anestrous bitches. In normal bitches, the serum progesterone level rises above 2 ng/mL within 2 months after estrus; progesterone level below 2 ng/mL for 6–8 months definitely indicates prolonged anestrus. In addition, vaginal epithelial cells should be monitored for any alterations with weekly vaginal smears.

Routine blood and urine tests (complete blood count, biochemical analysis) and thyroid function tests should be performed following general examination in anestrous bitches. Progesterone measurement and ultrasonographic examination of the ovaries should be carried out whenever luteal cyst or tumor of the ovaries is suspected. Karyotype analysis should be performed if developmental anomalies of the genital organs are suspected (such as hermaphroditism) [5].

Treatment of primary or secondary anestrus is targeted towards its etiology. No treatment is available if a bitch has undergone ovariectomy or ovariohysterectomy or in the presence of congenital sex development or differentiation anomalies, ovarian aplasia, or autoimmune oophoritis. Ovariohysterectomy is recommended for autoimmune oophoritis. Hormone replacement therapy may be administered in bitches determined to have hypothyroidism-related anestrus [22].

Estrous stimulation may be performed in anestrous bitches for which the underlying cause cannot be detected. Synthetic estrogens (diethylstilbestrol), dopamine agonists (bromocriptine and cabergoline), GnRH agonists (lutrelin, buserelin, fertirelin, deslorelin, leuprolide), or exogenous gonadotropins (LH, FSH, hCG, PMSG) may be administered for this purpose [23, 24].

Dopamine agonists are effective for stimulating fertile estrus in most bitches; however, prolonged use may be required. Bromocriptine (Parlodel, Gynodel) or cabergoline (Galastop, Dostinex) is used for this purpose. Bromocriptine is less preferred, as it causes vomiting and requires lengthy periods of administration in order to stimulate estrus. Although more expensive than bromocriptine, cabergoline may induce estruses effectively and safely with fewer side effects. Cabergoline is usually recommended at a dose of 5 μg/kg daily via the peroral route until 3–8 days after proestrus has begun [23–25].

Using GnRH and its analogs may also induce estruses. However, prolonged use for 8 days or longer or long-acting analogs such as lutrelin, deslorelin, or leuprolide is needed to induce fertile estrus. It is impractical to induce estrus by using short-acting natural GnRH or GnRH agonists, because GnRH should be given as a pulsatile continuous infusion at a dose of 0.2–0.4 μg/kg every 90 min via the intravenous or the subcutaneous route for 3–9 days. This requires 3–9 days of hospitalization and the availability of a pulsatile infusion pump. Long-acting formulations of GnRH analogs such as lutrelin, deslorelin, and leuprolide have been quite successful when implanted subcutaneously or submucosally. Deslorelin-containing implants (Suprelorin®, 4.7 mg deslorelin) are used most frequently for this purpose [23–27].

Hypophyseal (FSH and LH) and chorionic (PMSG and hCG) gonadotropins are also used for inducing estruses. It has been found that chorionic gonadotropins are more successful than hypophyseal gonadotropins in bitches. Although various protocols have been attempted, successful results have been reported with 20 IU/kg/d PMSG applied subcutaneously for 5 days, with 500 IU intramuscular hCG on day 5. PG600, a preparation containing PMSG and hCG first produced for pigs (80 IU PMSG and 40 IU hCG/mL), is quite effective for inducing estruses [8, 28].

1.3. Recurrent estrus (shortened interestrous intervals or polyestrous)

The mean duration between estruses is 7 months (4–13 months) in bitches, and the long part of the cycle (2–10 months) comprises a mandatory anestrus phase following diestrus. In the anestrous phase, the uterus enters the involution process and the endometrium is regenerated. Anestrus shorter than 2 months naturally results in repetition of estrus at 4-month or shorter intervals, which is defined as "recurrent estrous." It should be kept in mind that the estrous period is shorter in breeds such as the *German Shepherd Dog, Rottweiler, Basset Hound, Cocker Spaniel,* and *Labrador Retriever*. Fertility decreases in bitches with recurrent estrus [29] from two major causes: (1) overstimulation of the ovaries due to the formation of follicular cysts and granulose cell tumors and (2) a premature decrease of progesterone during diestrus [21].

The period between estruses may be prolonged up to 6 months by using a weak androgen, mibolerone (Cheque Drops), recommended at a dose of 30–180 μg per day to suppress estruses [1, 30].

Furthermore, if infertility resulted from short estrous period, bitches are reported to return to normal fertility in the following cycles when estrus is suppressed with synthetic progestin administration. For this purpose, 2 mg/kg per day megestrol acetate or 0.5 mg/kg per day of chlormadinone acetate may be administered for 8 days, so administration should begin within a maximum of 3 days following the beginning of proestrus [29].

1.4. Prolonged interestrous interval

Prolonged interestrous interval is defined as an interestrous period longer than 12 months. While estrus repeats 12–13 months after the previous estrus (prolonged interestrous interval) in some adult bitches, some are not observed to experience estrus again for a long time (secondary anestrus). It should be kept in mind that the interestrous period is longer in breeds such as the *Basenji* and *Tibetan Mastiff* compared to other breeds. Secondary anestrus may result from hypothyroidism, administrations of hyperadrenocortisolism, hyperprolactinemia, progesterone-secreting ovarian cysts, progestogen, androgenic or anabolic steroid substances, systemic diseases, poor nutrition, or housing in an inappropriate environment. Thyroid function should be evaluated first in bitches with prolonged interestrous intervals. Cycles typically return to normal when the underlying cause is treated [5].

http://dx.doi.org/10.5772/64648

1.5. Prolonged estrus (persistent estrus)

The mean duration of estrus is 9 days in an adult dog, which may sometimes be prolonged up to 3 weeks. Estrus of longer than 21 days with the absence of ovulation at the end of this long period is defined as prolonged estrus. Prolonged estrus is related to persistent and elevated estrogen levels, which remain continuously high during the estrous phase of the cycle. This disorder is encountered frequently in younger bitches, especially during the second cycle [1].

The most important clinical signs of the continuation of estrus include cornification in vaginal epithelial cells, continuation of the desire for copulation, vulvar edema and swelling, and hyperemia in vaginal mucosa for longer than 21 days. The serum progesterone level is low, while estrogen level is high.

Persistent estruses are usually related to an estrogen-releasing source, which may be an anovulatory follicle, follicular cysts, or functional ovarian tumors (granulose cell tumors). The follicles that develop in bitches receiving exogenous gonadotropins in order to experience estrus may sometimes lead to prolonged estruses. Exogenous estrogen administration for the treatment of urinary incontinence or vaginitis, hormone replacement therapy, and prevention of undesired pregnancy may also cause persistent estruses. Persistent estrus may also develop alongside tumors of the hypophysis or the hypothalamus or in a hepatic disease defined as portosystemic shunting in which an abnormal vascular junction is formed between the hepatic portal vein and the systemic circulation. In these cases, metabolism of estrogen in the liver is impaired.

Determination that 90% or more of the cells in a vaginal smear specimen are permanently cornified on cytological examination and nondetection of the normal increase in serum progesterone levels (remaining within the preovulatory range <2 ng/mL) indicate prolonged estrus. Detection of serum estrogen level is not a reliable method for diagnosis [1].

The first step in treatment should include a determination of the source of estrogen causing prolonged estrus. For this purpose, the ovaries should be examined ultrasonographically for the presence of abnormal structures (e.g., ovarian cyst, granulose cell tumor); if ovarian structures cannot be identified by ultrasonographic examination, exploratory laparotomy should be performed, followed by biopsy, if required.

Follicles or follicular cysts causing prolonged estrus may heal spontaneously. If estrus is determined to last longer than 3 weeks, interventions are recommended in order to prevent bone marrow hypoplasia and/or pyometra. Treatment options should match the pet owner's expectations regarding having a puppy; if a puppy is not expected, ovariohysterectomy is the best option.

Ovulation or luteinization may be obtained by GnRH or hCG injections into follicles if the pet owner wants a puppy. hCG administration at a dose of 22 IU/kg via the intramuscular route for 3 days and GnRH (gonadorelin) administration at a dose of 10 μg/kg via the intramuscular route for 3 days are recommended. Copulation is not recommended, as the target of these applications is not the induction of ovulation but rather the termination of the signs of prolonged estrus [1]. Ovariohysterectomy is inevitable in cases in which no response is obtained from medical applications and in prolonged estrus cases due to ovarian tumors [5].

Megestrol acetate (Ovaban, Ovarid) may be applied to reduce the signs of prolonged estrus. Low doses of megestrol acetate are recommended via the peroral route for 2 weeks. A dose of 0.1 mg/kg is proper for the first week, and a dose of 0.05 mg/kg is proper for the second week. Although progesterone treatment with megestrol acetate is effective in bitches with persistent estrus, there is the potential to trigger the development of cystic endometrial hyperplasia. Therefore, the treatment is contraindicated for bitches to be later considered for copulation. In general, ovariohysterectomy is performed within 3 weeks after treatment with progesterone in bitches with persistent estrus.

Bone-marrow suppression related to long-term estrogen toxicity may develop in bitches with persistent estrus. Nonregenerative anemia and thrombocytopenia are observed in such bitches [31]. Therefore, erythropoiesis-stimulating agents, such as synthetic erythropoietin, darbepoetin, granulocyte-colony stimulating factor, and granulocyte-macrophage colony-stimulating factor, may be used beside proper antibiotic and blood products in bitches that have anemia due to bone-marrow suppression. In addition, lithium [30, 32], synthetic anabolic steroids such as nandrolone decanoate (Deca-Durabolin), or a dihydrotestosterone derivative such as stanozolol (Winstrol) can be quite useful [33, 34].

1.6. Prolonged proestrus (persistent proestrus)

Prolonged proestrus is defined as a proestrous phase that is not followed by an estrous phase and that lasts 3 weeks or longer. In bitches with prolonged proestrus, estrus and ovulation do not occur, as the estrogen level insufficiently increases during the proestrous phase.

Prolonged hemorrhagic vaginal discharge, cornified cells higher than 50–90% on examination of vaginal smear, and serum progesterone level remaining below 2 ng/mL indicate prolonged proestrus. Treatment principles are similar to those for prolonged estrus.

1.7. Split estrus

Split estrus is a disorder in which no or quite short estrous signs develop despite the presence of proestrous signs. In this situation, pregnancy usually does not develop even if copulation occurs; the bitch is observed to enter proestrus again within 3–4 weeks. In these bitches, the next cycle is usually a normal ovulatory cycle.

Split estrus is usually seen in young bitches that have shown first estrus. However, continuous or frequent split estruses should suggest chronic premature luteolysis or hypothyroidism. Split estrus may be confused with recurrent estrus (short interestrous interval). Ovulation will not develop in dogs showing split estrus but without the typical progesterone elevation. The condition usually recovers spontaneously [1, 7].

1.8. Anovulatory cycle

A serum progesterone level not exceeding 2 ng/mL despite cytological estrous signs is defined as anovulation. Although the cell type in vaginal cytology is noncornified, diestrous-specific progesterone elevation does not occur, and the bitch enters anestrus.

http://dx.doi.org/10.5772/64648

The most typical signs are low serum progesterone levels and the absence of ovulation during the days after copulation in a bitch showing proestrous and estrous signs. Its incidence is about 1%. The following ovulatory cycles were observed to be normal in 45% of bitches that had an anovulatory cycle [1, 21, 35].

Thus, treatment is usually not required in bitches with an anovulatory cycle. hCG or GnRH may be administered, if desired; however, their application carries the potential to trigger pyometra [1, 21].

2. Conclusion

There are many factors leading to abnormalities in the sexual cycle of bitches. A decent anamnesis is required to find out the causes of these abnormalities. Supporting the anamnesis information by clinical and laboratory examinations is of importance for the accuracy of the diagnosis. Vaginal cytology among the diagnostic methods should be used and interpreted efficiently. Accuracy of the diagnosis forms the first step of an effective treatment. Uses of hormones, particularly gonadotropins, come into prominence in the treatment of sexual abnormalities in these animals.

Author details

Ali Risvanli*, Halis Ocal and Cahit Kalkan

*Address all correspondence to: arisvanli@firat.edu.tr

Department of Obstetrics and Gynecology, Faculty of Veterinary Medicine, University of Firat, Elazig, Turkey

References

[1] Meyers-Wallen VN. Unusual and abnormal canine estrous cycles. Theriogenology 2007;68:1205–1210.

[2] Johnston SD. Clinical approach to infertility in bitches with primary anestrus. Vet Clin North Am Small Anim Pract 1991;21:421–425.

[3] Kalkan C, Ocal H. Reproductive physiology. In: Kaymaz M, Findik M, Risvanli A, Koker A, editors. Obstetric and Gynecology in Bitches and Queens. 1st ed. Malatya: Medipres; 2013. p. 27–62.

[4] Holyoak GR, Makloski C, Morgan GL. Abortion, abnormal estrous cycle, and infertility. In: Lorenz MD, Neer TM, DeMars P, editors. Small Animal Medical Diagnosis. 3rd ed. USA: Wiley-Blackwell; 2013. p. 337–354.

[5] Nak D, Kasikci G. Infertilty. In: Kaymaz M, Findik M, Risvanli A, Koker A, editors. Obstetric and Gynecology in Bitches and Queens. 1st ed. Malatya: Medipres; 2013. p. 223–273.

[6] Feldman EC, Nelson RW. Canine and Feline Endocrinology and Reproduction. 3rd ed. USA: Saunders; 2004.

[7] Grundy SA, Feldman E, Davidson A. Evaluation of infertility in the bitch. Clin Tech Small Anim Pract 2002;17:108–115.

[8] Kustritz MVR. The Dog Breeder's Guide to Successful Breeding and Health Management. Philadelphia: Saunder Company; 2006.

[9] Place NJ, Hansen BS, Cheraskin JL, Cudney SE, Flanders JA, Newmark AD, Barry B, Scarlett JM. Measurement of serum anti-Müllerian hormone concentration in female dogs and cats before and after ovariohysterectomy. J Vet Diagn Invest 2011;23:524–527.

[10] Turna Yilmaz O, Toydemir TS, Kirsan I, Gunay Ucmak Z, Caliskan Karacam E. Anti-Mullerian hormone as a diagnostic tool for ovarian remnant syndrome in bitches. Vet Res Commun 2015;39:159–162.

[11] Lyle SK. Disorders of sexual development in the dog and cat. Theriogenology 2007;68:338 343.

[12] Poth T, Breuer W, Walter B, Hecht W, Hermanns W. Disorders of sex development in the dog-adoption of a new nomenclature and reclassification of reported cases. Anim Reprod Sci 2010;121:197–207.

[13] Meyers-Wallen VN. Gonadal and sex differentiation abnormalities of dogs and cats. Sex Dev 2012;6:46–60.

[14] Christensen BW. Disorders of sexual development in dogs and cats. Vet Clin North Am Small Anim Pract 2012;42:515–526.

[15] Marino G, Rizzo S, Quartuccio M, Macrì F, Pagano G, Taormina A, Cristarella S, Zanghì A. Deslorelin implants in pre-pubertal female dogs: short- and long-term effects on the genital tract. Reprod Domest Anim 2014;49:297–301.

[16] Liu Y, Tian Y, Zhao X, Jiang S, Li F, Zhang Y, Zhang X, Li Y, Zhou J, Fang F. Immunization of dogs with recombinant GnRH-1 suppresses the development of reproductive function. Theriogenology 2015;83:314–319.

[17] Chastain CB. Canine pseudohypothyroidism and covert hypothyroidism. Probl Vet Med 1990;2:693–716.

[18] Johnson CA. Thyroid issues in reproduction. Clin Tech Small Anim Pract 2002;17:129–132.

[19] Scott-Moncrieff JC. Clinical signs and concurrent diseases of hypothyroidism in dogs and cats. Vet Clin North Am Small Anim Pract 2007;37:709–722.

[20] Sontas BH, Schwendenwein I, Schäfer-Somi S. Primary anestrus due to dietary hyperthyroidism in a miniature pinscher bitch. Can Vet J 2014;55:781–785.

[21] Fontbonne, A. Infertility in bitches and queen: recent advances. Rev Bras Reprod Anim 2011;35:202–209.

[22] Simpson C, Devi JL, Whittem T. Bioavailability of two L-thyroxine formulations after oral administration to healthy dogs. Aust Vet J 2013;91:83–88.

[23] Kutzler MA. Estrus induction and synchronization in canids and felids. Theriogenology 2007;68:354–374.

[24] Wiebe VJ, Howard JP. Pharmacologic advances in canine and feline reproduction. Top Companion Anim M 2009;24:71–99.

[25] Gobello C, Castex G, Corrada Y. Use of cabergoline to treat primary and secondary anestrus in dogs. JAVMA 2002;220:1653–1654.

[26] Fontaine E, Mir F, Vannier F, Gérardin A, Albouy M, Navarro C, Fontbonne A. Induction of fertile oestrus in the bitch using Deslorelin, a GnRH agonist. Theriogenology 2011;76:1561–1566.

[27] Walter B, Otzdorff C, Brugger N, Braun J. Estrus induction in Beagle bitches with the GnRH-agonist implant containing 4.7 mg Deslorelin. Theriogenology 2011;75:1125–1129.

[28] Popescu MC, Nicorescu V, Codreanu I, Crivineanu M. General considerations according to pituitary versus placental gonadotrophins activities in bitch. Vet Med 2012;58:302–305.

[29] Wanke MM, Loza ME, Rebuelto M. Progestin treatment for infertility in bitches with short interestrus interval. Theriogenology 2006;66:1579–1582.

[30] Tilley LP, Francis WK, Smith JR. Blackwell's Five-Minute Veterinary Consult: Canine and Feline. 6th ed. USA: Wiley-Blackwell; 2015.

[31] Sontas HB, Dokuzeylul B, Turna O, Ekici H. Estrogen-induced myelotoxicity in dogs: A review. Can Vet J 2009;50:1054–1058.

[32] Chalhoub S, Langston CE, Farrelly J. The use of darbepoetin to stimulate erythropoiesis in anemia of chronic kidney disease in cats: 25 cases. J Vet Intern Med 2012;26:363–369.

[33] Shimoda K, Shide K, Kamezaki K, Okamura T, Harada N, Kinukawa N, Ohyashiki K, Niho Y, Mizoguchi H, Omine M, Ozawa K, Haradaa M. The effect of anabolic steroids

on anemia in myelofibrosis with myeloid metaplasia: retrospective analysis of 39 patients in Japan. Int J Hematol 2007;85:338–343.

[34] De Bosschere H, Deprest C. Estrogen-induced pancytopenia due to a sertoli cell tumor in a cryptorchid Beauceron. Vlaams Diergen Tijds 2010;79:280–284.

[35] Arbeiter K. Anovulatory ovarian cycles in dogs. J Reprod Fertil 1993;47:453–456.

Chapter 7

Pursuing Alternative Strategies for Healthier Medical Contraception in Dogs

Rita Payan-Carreira , Paulo Borges and
Alain Fontbonne

Additional information is available at the end of the chapter

http://dx.doi.org/10.5772/65331

Abstract

Although extensively used in the control of the reproductive cycles in either the domestic or feral dogs as well as in wild carnivores, medical progestin-based contraception still raises concerns to the veterinary practitioner and owners on its safety and efficiency. These concerns endorsed, in last decades, the research in the development of new alternatives for effective, reversible, and safe contraceptive methods for carnivores, mainly pursuing a larger-scale control of canine reproduction and the development of products with few side effects. Nowadays, the medical contraceptives often intend to master, in a reversible way, the reproductive cycle in genetically valuable dogs, which presumes that they would be active for short periods of time and ought to safeguard the animal fertility. However, hormonal contraceptives are also used worldwide to control the reproductive activity in either domestic or feral cats, for long-term treatments, because of a pretended short-term economic interest. Progestogens are the most frequently used hormonal contraceptive in carnivores. They are rather easy to obtain across the globe and relatively cheap; they have diverse drug presentations, allowing their use independently of the veterinary assistance, and are effective in preventing pregnancy. Still a significant number of undesirable health side effects are attributed to progestins when employed with some chronicity, when applied in older animals or even when misused. In the past two decades, several new approaches to managing dog reproduction were proposed to avoid progestins. However, their efficiency and cost are still to be proven as a viable alternative around the world. This chapter aims to review the medical methods available as alternative to the progestins in canine contraception, addressing particularly the future perspectives, opportunities, and limitations linked to currently available substitutes, based on our practice. This information can be of utmost interest to students, clinicians or colonies' technicians.

Keywords: medical contraception, estrus suppression, GnRH antagonists, GnRH agonist, immunocontraception, dogs

1. Introduction

The control of the reproductive activity in dogs is a main issue in today's societies. Overpopulation of stray and feral dogs is an universal problem and raises public health concerns related to the increased risk of zoonotic diseases and compromise of the environmental health that may foster the permanence of diseases in other species or the degradation of ecosystems [1], as well as to the social and urban problems associated with animal-human conflicts.

At the start of this chapter, the authors would also strengthen that the best measure for a safe and definitive measure for the suppression of reproductive cycles in animals not intended for future breeding is ovariectomy (in prepubertal or young postpubertal female dogs) or ovariohysterectomy [2–4]. These procedures also have a protective effect on the incidence of uterine and mammary diseases, if performed early in the animal life. However, spaying of young animals, whether male or female dogs, is not an universal option; in some countries, it is described as an unnecessary surgical insult to the dog, and therefore must be performed only under medical indication. In other, the surgery is rejected for economic reasons, and medical contraception is regarded as a more economical alternative to surgery. The therapeutic options for mastering the companion animal reproduction substantially improved in recent decades. Also, the nonsurgical contraceptive treatments drew the attention of the industry and researchers, allowing the recent introduction of new methods in the clinical practice targeting safer long- and short-term suspension of gonadal activity in dogs. The information on the available methods alternative to progestins needs to be discussed and their advantages and disadvantages have been reviewed to disseminate basic information of their use and ease the introduction of safer products in canine medicine [5].

The suppression of canine reproductive activity includes: (1) the sterilization of both male and female, as a form to control either the ownerless or free-roaming or community-owned dogs as well as privately owned, confined or roaming dogs; and (2) the canine contraception for the periodic suspension of ovarian activity, which can also be applied on the male counterpart. This chapter focuses on the medical contraception in dogs. Moreover, as the manipulation of the ovarian activity in dogs, even if directed to its suspension, presumes the basic knowledge of the normal endocrinology of the estrous cycle of bitch, it will be reviewed, before discussing the nonsurgical contraceptive treatments available in dogs.

2. The canine estrous cycle

Dogs are spontaneous ovulators, monoestrous, and nonseasonal species [6, 7], despite that there are some reports on an increased incidence of estrus during winter and late spring [8]. However, the photoperiodic control of estrus has only been clearly shown in the Basenji [9].

In dogs, each cycle is separated from the following by an obligatory anestrus stage of variable duration [6]. The period between two consecutive proestruses is often named as interestrous interval. The number of cycles per year varies greatly among bitches, accounting the high variability recorded for the interestrous interval among this species. However, this variation is independent of the animal size. Although the pattern and regularity of the estrus activity may vary between breeds or even genetic lines, it is regular for each female. Some females show only one cycle per year, such as the Basenji and the Tibetan Mastiff, others show two to three cycles yearly. Thus, the physiological interestrous interval may vary between 4 and 12 months [6, 10]. For example, some lines of Rottweilers, German Sheepdogs, and Bernese Mountain dogs may show a 4-month interestrous interval, while some lines of Collies, Labradors or Teckels may present an interval of 7–8 months. Usually, the reported average is 6–7 months [6, 7]. Therefore, it is important to collect the information on the regularity of the estrous cycle in a particular bitch when designing a contraceptive medical protocol, as this may require the adjustment of the administration schedule.

Age at puberty in dogs is mostly affected by size, although nutritional or social cues may also modulate it. In female dogs, puberty occurs when the animal reaches 70–80% of its mature body weight [11].

Classically, the canine estrous cycle is divided into four stages (**Figure 1**) that recur at regular intervals [6–8]. The length of each stage of the canine estrous cycle varies individually, with exception of the diestrus that is fairly constant whether pregnancy occurs or not.

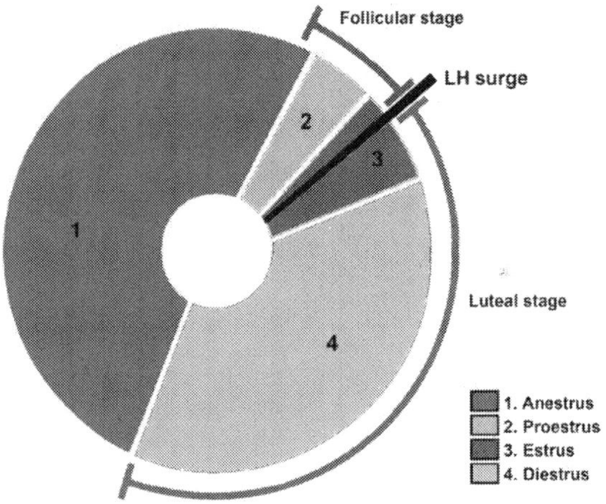

Figure 1. Schematic representation of the standard canine estrous cycle. Adapted with permission from [12].

Proestrus represents the first signs of reproductive activity; following the rapid follicular development of follicles in the ovary, which determines a rapid increase in the suprabasal estrogen levels, the female presents the external clinical signs associated with heat, such as swollen vulva, serous-hemorrhagic vulvar discharge, and increased restlessness and attraction of male [6–8]. The average length of proestrus is 9 days, but in fact it may range between 3 and 21 days [7]. The transition from proestrus to estrus is feebly detected on the basis of external or behavioral features, and so the two stages are usually grouped under the designation of "heat". Estrus is a transitional stage. Owing to the preovulatory luteinization of the granulosa cells in the growing follicles (progesterone levels increase above 2 ng/mL from LH peak onwards), early in this stage occurs a shift in the steroid environment, which changes from peaking estrogens to the progesterone dominance that will be maintained throughout diestrus. These changes induce a decrease in the amount of vaginal discharge, which also becomes more mucous and less hemorrhagic. In this stage, the female search more actively and the contact with the male and allows mating. Estrus lasts in average 9 days, but individual variations account for a range of 3–21 days. The LH surge occurs usually 24 h after the onset of estrus, but ovulation will take another 1.5–2.5 days to occur. The ovulation product is an immature oocyte that needs an additional period of 2–3 days for tubal maturation before fertilization [6].

Diestrus represents a prolonged luteal stage, similar in length whether or not pregnancy occurs. The decrease of progesterone levels below 1–2 ng/mL is often used to delimit the end of diestrus, as externally no clinical signs allow establishing the limits between diestrus and anestrus. Still, some bitches may present a residual mucous vaginal discharge or mammary development during diestrus that are absent in anestrus. The mean length of nongestational diestrus is 60 ± 15 days, while the gestational diestrus lasts for 63 ± 1 days. The levels of progesterone are already high at the diestrus onset; the peak is maintained for almost half the stage and gradually decreases by the end of this stage; the progesterone decline is more abrupt in gestational cycles than in nonpregnant ones [6, 8].

In anestrus, the sex steroids are maintained in basal values, except in the last third of the stage, when the initial development of a wave of follicles in the ovaries occurs, thereby inducing a small increase in estrogens [13, 14]. Although this stage is often considered a time of reproductive quiescence, in fact in the uterus an important remodeling and repairing of the endometrium occurs, which is of upmost importance to the bitch fertility [13]. The length of this stage is the most variable in dogs, despite the reported average length of 18–20 weeks. The minimum length of anestrus is 7 weeks after the progesterone drop, but it can reach up to 10 months [6, 7]. It is important to remember, however, that the duration of the anestrus may be modulated by external environmental factors. An anestrous bitch can be stimulated to resume proestrus when in close proximity to a bitch in estrus [7]. This in fact contributes to the synchronization of estrous cycles often observed when bitches are housed together.

3. Nonsurgical contraceptive options

Lately, in most developed countries, a variety of possible contraceptive methods became available. Thus, allowing the owner of a female dog or the practitioner to adopt the plan that best suits a particular individual or situation, by weighting the owner's aims for treatment, the physiological condition of the bitch and the expected side effects of the selected method. However, in a significant number of countries, progestins may remain the option of choice, mainly due to economic constraints.

The target population or individual is an important parameter, along with the requested period of estrus suppression. The categories of dogs (owned dogs vs. community owned or ownerless, free-roaming dogs) [15] often dictate the selection, effectiveness, and feasibility of the contraceptive method, because they determine the regularity of the drug administration, the easiness of administration and the ability to survey the animal during treatment. Therefore, temporary contraception is most suitable for privately owned dogs, while for community owned or ownerless dogs, sterilization (surgical or chemical) remains the best decision.

Below the available therapeutic options for contraception of individual bitches will be discussed.

3.1. Progestins

Progestins (synthetic progesterone-like compounds, also known as progestogens) remained for long time the unique available medical contraceptive option in dogs. Intended for a short-term estrus suspension or postponement, chronic treatments longer than 2 years usually increase the negative effects that these drugs exert over the endocrine axis and the reproductive tract of the bitch.

Progestins are widely used, although these substances present major detrimental side effects in dogs, whose sensitivity to prolonged progesterone is high and predispose to uterine and mammary diseases [16]. This is a major drawback for the progestins use, particularly in chronic administration protocols or whenever the administration timing and doses are not followed adequately.

Progestins place the female under a prolonged artificial luteal stage. A constant supply of progestins causes the gonadotropin-releasing hormone (GnRH) down-regulation, which in turn depress the follicle-stimulating hormone (FSH) and luteinizing hormone (LH) secretion, therefore suspending the follicular development in the ovaries. Also, progestins change the viscosity of the tubal and uterine secretions, and reduce the motility in the reproductive tract, compromising the transport of gametes and eggs and ruining the receptivity for a potential embryo [17]. Progestin actions are directly related to their side effects, as summarized in **Figure 2**.

The importance of the reactions to progestins administration increases the risk for a disease. This risk vary with: (1) the drug, its dose, the via of administration, and the frequency/length of the treatment; (2) the stage of the estrous cycle at the onset of treatment; and (3) some individual variations in sensitivity to the progestins that may be related to the age of the female (females older than 5 years are poor candidates for progestin treatment), existing pathologies or predispositions. Therefore, before starting any progestin treatment, the female should be subjected a thorough clinical examination. A careful and detailed anamnesis will allow to ascertain whether the female is postpubertal and in anestrus, whether the female is intended to breed within a 2-years period or eventually to understand if the owner will discontinue the treatment after a period of 18–24 months of progestin administration [4, 18] and willing to allow a full-term pregnancy to minimize the deleterious effects of progestins.

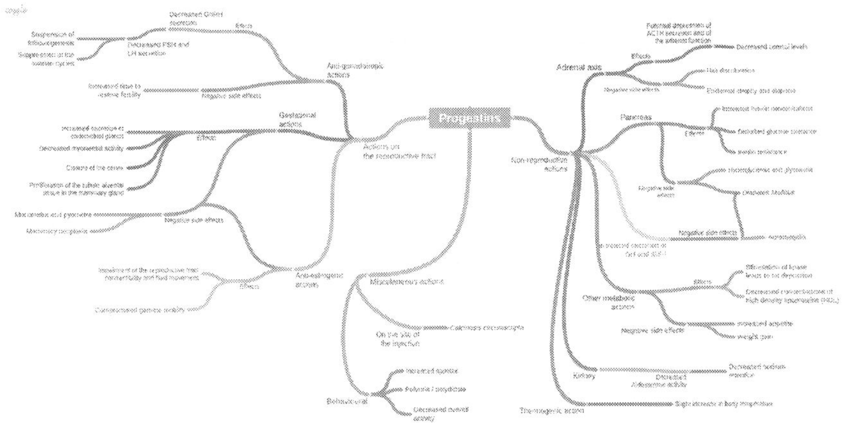

Figure 2. Summary of the main actions of progestins and the corresponding detrimental side effects in dogs.

The physical and reproductive examination will allow excluding pregnancy and obesity or any clinical evidence of hepatic, uterine, mammary, or metabolic disease, which might be exacerbated by the progestin treatment [17]. A female that is already in the proestrus stage is also a poor candidate for a progestin treatment, despite that these drugs might be (exceptionally) administered in the first day of heat [18]. The onset of treatment at the beginning of the follicular stage will increase the risks of pyometra [17] or mammary diseases.

There are several available progestins in the veterinary market, resulting from the research for less harmful products for dogs. However, the progestogen formulations approved for dogs may vary with the country. According to the chemical composition, progestins present different antigonadotrophic, gestagenic, and antiestrogenic properties that also define different risk potential. Older drugs (for instance, medroxyprogesterone acetate or megestrol acetate) usually possess stronger gestagenic actions and therefore more powerful negative side effects on the uterus and mammary glands than recent generation products, like proligestone.

In general, the particularities of the canine estrous cycle demand a different schedule for administration of oral progestins aiming the suspension of heat or the suppression of the estrous cycle. Thereby, in this species, the administration of progestins is more frequent as an injectable formula. However, it is important to know the length of the interestrous interval of a particular bitch and to adjust the subcutaneous administration of sequential progestins to avoid the breakdown effect (i.e., failure to control the cycle and by consequence the bitch enters in heat during the interval between administrations). Moreover, it should also be important to adjust the dose to the actual body weight, particularly in females of large and giant breeds.

Progestins available in the market for dogs include:

- **Medroxyprogesterone acetate** (MPA) — one of the first progestins used in dogs, it possesses high-antigonadotrophic action and a high-gestagenic action [17, 19]. The veterinary formulation may present different names according to the country (i.e., Supprestral® or Perlutex®) and often is marketed for both oral and intramuscular administration. However, in several countries, the human medicine injectable formulation is often used due to its reduced cost. However, the putative MPA negative side effects are high, particularly when the correct doses are not followed (2 mg/kg every 3–4 months or 3–5 mg/kg at 5–6 months interval, for a maximal of four repetitive administrations). Individual variations can be observed in the response to estrus suppression in some females. Consequently, its use should be avoided in valuable females [19]. Close monitoring of treated females is crucial for an early detection of MPA side effects.

- **Megestrol acetate** (MA)—a short-acting progestin that shows less negative side effects than MPA and therefore it is most commonly used for temporary estrus suppression [17, 19]. In dogs, MA is often used in oral presentations (such as Ovarid®/Ovaban® or Pilucalm®, although the name of the product may vary with the country), in a daily administration at the doses of 0.5 mg/kg, for 32 to a maximum of 40 days, starting 14–30 days before the expected heat. It may be repeated at 6 months of interval, for a total of two administration cycles.

- **Delmadinone** (DMA) and **Chlormadinone acetate** (CMA) —the two progestins with a limited spread in the global veterinary market, and as consequence, information on its application in dogs is scant. DMA and CMA are usually used as antitestosterone in male dogs for the treatment of behavioral or prostate problems. The available DMA (Tardak®) doses for dogs are 1–2 mg/kg, to be injected subcutaneously, every 6 months. All treatments should start in anestrus [20]. The dose for the CMA injectable formulation in dogs is 3 mg/kg, subcutaneously, to be administered 1 month before the expected heat and repeated in a 6 months of interval [21]. When treatments start in anestrus, the product is said to be safer than MPA or MA, although it presents a similar increase in body weight.

- **Proligestone** (PRG)—a new generation progestin, possesses higher antigonadotrophic action but feebler gestational and antiestrogenic effects in comparison to MPA and MA [4, 19]. Therefore, it presents a lower incidence of mammary or uterine disease, compared to the other progestins. PRG (Delvosteron®/Covinan®, Intervet) is recommended at a dose of 10–33 mg/kg, starting 2 weeks before the expected season. The second administration should be repeated after 3 months, and afterwards at a 5-month interval. PRG is considered a safer drug than older progestins and reported to have minimal side effects [4, 19].

3.2. Androgens

Natural or synthetic androgens, like progestins, induce a down-regulation of the hypothalamus-pituitary-gonadal function. Thus, they can be used for heat suppression. Androgens were often used in racing Greyhounds to avoid estrus in training or racing females [22]. The anabolic side effects and the virilisation (clitoral hypertrophy and anal gland inspissation) associated with androgen administration [2, 20] were tolerated in Greyhounds, but may raise ethical concerns in the practice. Additional side effects described for androgen contraceptive treatments include vaginitis and urinary incontinence [18]. However, a major concern on the use of androgens for contraception was related to the induction of a rather prolonged anestrus (ranging between 1 and 2 years) [20]. Thereby their use would compromise the use of such females for breeding after the interruption of treatment and dismissal from the racetrack [22]. Anyway, these drawbacks compromised the use of androgens in current clinical practice [7].

Few studies exist on the use of androgens as estrus suppressor in dogs. The oral administration of androgens is described for methyl testosterone (25 mg/dog, once a week [2]), orandrone (0.5 mg/kg, daily [20]), methyl testosterone associated with ethinylestradiol (7 mg/kg, daily, for 5–10 days.[20]), and mibolerone, a synthetic weak androgen (Cheque Drops, at a dose of 30–180 mcg/day, starting 30 days before the onset of heat; it should not be administered for more than 2 years [7]). Injectable androgen therapeutic options include the intramuscular administration of testosterone cypionate (1 mg/kg, every 2 weeks [22]), testosterone phenylpropionate (110 mg/dog, weekly or alternatively at 0.5–1 mg/kg, every 7–10 days [2] and [20], respectively) or a composition of four different testosterone esters (Durateston®, at a dose of 2.5–5 mg/kg every 6 months [20]).

3.3. GnRH agonists

The GnRH agonists are intended to suppress the pulsatility of GnRH and indirectly that of LH and FSH [23]; consequently, these compounds depress the follicular activity in the ovaries and reduce to baseline the secretion of sexual steroids. Shortly, they reset the bitch into a prepubertal stage or in anestrus. By suppressing the secretion of progesterone, these substances will minimize the risk for uterine or mammary diseases found in the progestins contraception. Furthermore, they are beneficial in controlling unwelcome sexual behavior associated with the female season and animal aggressiveness. However, these products are usually expensive, which limits their wider utilization and reduces their competitiveness compared to progestins.

The available commercial products include azagly-nafarelin implant (Gonazon®), deslorelin acetate implants (Suprelorin® or Ovuplant®), and leuprolide acetate implant (Lupron Depot®), although the application of buserelin (50 μg, single administration) may also accomplish the GnRH downregulation. However, the length of the induced anestrus is less precise and the individual variation in the response is higher.

GnRH agonists are presented today as a subcutaneous implant, allowing a prolonged controlled release of the drug into the system. Implants are injected subcutaneously at the interscapular or the postumbilical area [24]. This sustained exposition to the GnRH agonist would override the endogenous secretion of this hormone. The agonist acts at the level of the GnRH receptors [23], inducing the receptor downregulation, internalization and signal uncoupling [4], resulting in the termination of the signaling cascade triggered GnRH in cells. However, GnRH agonists act in a dual phase mechanism. Before acting as described before, the first effect of implants is to stimulate the pituitary axis ("flare-up" effect), triggering an initial release of FSH and LH [25], thus shortly originating a new season in treated bitches. The "flare-up" effect is usually observed within 1 month after the insertion of the implant [26]. This is, to date, the most significant drawback identified for the use of GnRH agonists in dog contraception: to first induce estrus before preventing it [27].

The flare-up effect is more frequent when treatment starts in late anestrus [28, 29], compared to any other stage of the canine cycle. Whether or not it may be decreased when the female is implanted in diestrus or given exogenous progestogen [30–32] is still controversial [33, 34].

Additional adverse side effects reported following the insertion of GnRH agonist implants include persistent estrus associated with the formation of ovarian cysts due to anovulation [35], galactorrhoea, metropathies, vomiting, cystitis, and allergic reactions [36]. These effects occur with variable incidence suggesting the existence of individual idiosyncratic factors. These may also comprise preexisting problems that remained clinically undiagnosed. Consequently, before injection of the implant, the female should be subjected to a thorough clinical and reproductive examination to exclude any ovarian and uterine pathology [24].

GnRH agonists can be used as short or long-term contraceptives in domestic carnivores [37, 38] and may be applied in prepubertal females to postpone the reproductive activity without apparent side effects to delay puberty [39, 40]. The effect of the implants may be fully reverted and fertility regained after the 12-month period of the implant action [23, 41]. However, in some cases, the implant effect can last up to 27 months [23, 32]. An anticipated removal of the implant would also permit the withdrawn of the effects over the hypothalamic-pituitary axis, and regain cyclicity earlier. Recently, it has stretched that Superlorin® presented no detrimental effects on the bitch fertility whether a short-term or the long-term treatment was implemented [24].

Figure 3 provides a summary analysis to GnRH agonists as contraceptives in dogs.

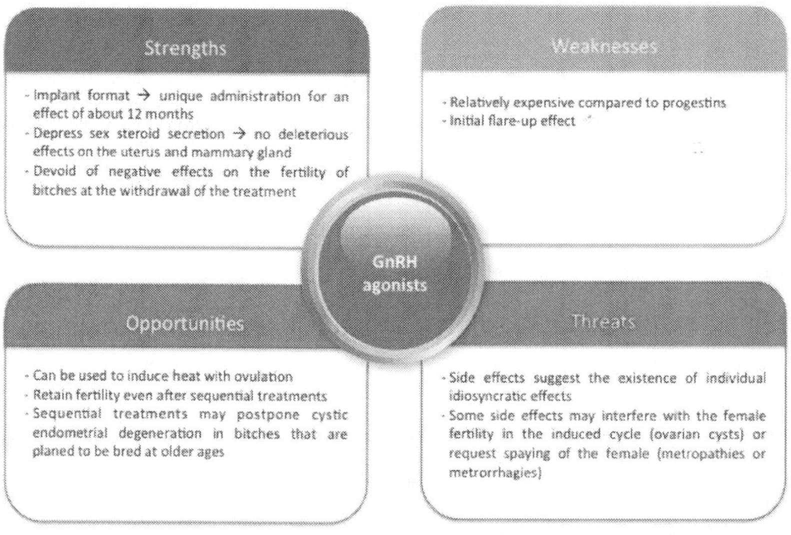

Figure 3. SWOT analysis to the use of GnRH agonists to control the bitch reproduction.

3.4. GnRH antagonists

The GnRH antagonists limit its action by a competitive block of the hormone receptors achieving the annulation of the effects of circulating GnRH. Consequently, the function of the hypothalamus-pituitary-gonadal axis is impaired. Long-term effects of GnRH antagonists also include the down-regulation of the GnRH receptors [4, 42]. Either way, follicular waves are suppressed and ovulation is compromised [19]. Several generations of peptides with GnRH antagonist activity have been tested in dogs, but their use is still limited all over the world, mainly because they present a rather low efficiency as contraceptives. Also, the first generation compounds showed several important side effects, derived from the need for higher doses of these peptides to reach the desired effect. Peptide GnRH antagonists act only for short-term estrus suppression [19, 43], which make them a poor agent when longer periods of contraception are foreseen. Therefore, its use is mostly restricted to the short-term contraception in show or work dogs [19].

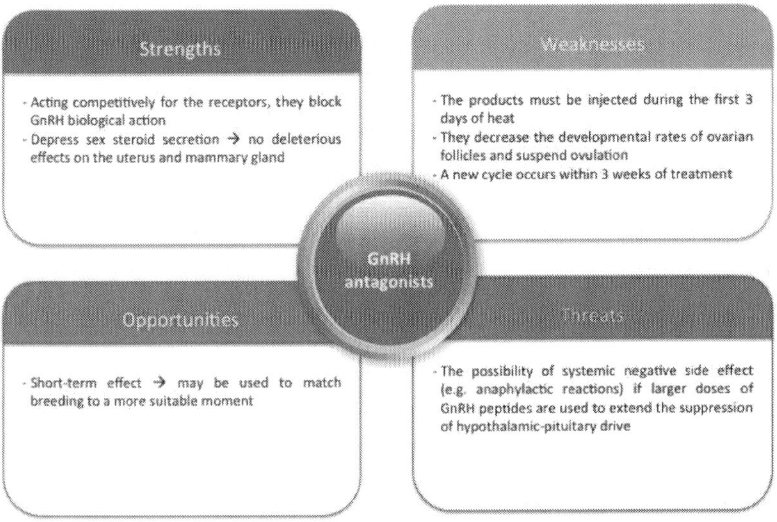

Figure 4. SWOT analysis to the use of GnRH antagonists as a contraceptive in dogs.

In dogs, the third generation GnRH antagonist, Acylin®, is the most used antagonist. It should be administered within the first 3 days of proestrus, subcutaneously, at a dose of 100 μg/kg. Suspension of the follicular stage is obtained, and ovulation inhibited; however, the bitch is expected to enter a new cycle within 3 weeks of treatment [4].

Companies are now exploiting some nonpeptide molecules as GnRH antagonists, alike those tested for humans, aiming to obtain the long-term release formulations that may be applied for the long-term contraception. However, to our best knowledge, no information is yet available on these molecules.

Figure 4 provides a summary analysis to the use of GnRH antagonists as contraceptives in dogs.

3.5. Immunocontraception

Immunocontraception is yearned for long estrus control, either for owned and free-roaming dogs. Classical targets in immunocontraception include the GnRH and LH along with their corresponding receptors, as well as the sperm or *Zona pellucida* (ZP) proteins [4, 44]. Vaccines are usually conjugated with various antigens to enhance the immune response against the target compound(s). Contraception would be maintained through regular boosting [45]. Either approach shows unsatisfactory results till present [4]. Particularly, the resumption of estrous cycles or fertility after withdrawal of the treatment is still a concern.

The use of vaccines targeting GnRH will induce suppression of the estrous cycle, while vaccines against LH will interfere with ovulation because the preovulatory LH-surge is suppressed, and progesterone secretion is also compromised. Depending on the vaccine used, the cycle may be suspended for periods ranging from 5 months to 5 years with a single administration. On the other hand, vaccines targeting sperm or *Zona pellucida* proteins will not disturb the normal estrous cycle of the female dog, but will inhibit egg-perm binding in the female genital tract and fertilization [15]. *Zona pellucida* vaccines do not succeed to induce infertility in dogs [4, 15].

GonaCon™ is a vaccine against GnRH. It was developed to control the reproduction in the wildlife population, in which infertility was achieved for a period of 1–4 years with a single vaccination [46]. However, its use in dogs is still controversial, as it seems that dogs present intense reactions at the site of injection, with formation of long-lasting abscesses and granulomas, due to a greater sensitivity to the adjuvant used in the vaccine. Also, there is no available data on the duration of estrus suppression [15]. Recently, in Mexico, a new formulation of GonaCon (with the adjuvant AdjuVac™) was tested in shelter dogs during a campaign for rabies and control of canine reproduction [47]. According to the data reported, the proportion of animals presenting abscesses at day 60 was lower than the expected from previous studies using different vaccine adjuvants, but it was still not devoid of other local side effects (like muscular atrophy) [47].

Another study, using an unidentified commercial vaccine against GnRH, administered twice at a 4-week interval, however, reported that none of the four animals used showed adverse reactions to vaccination, remaining clinically healthy for the length of the study (20 weeks) [48]. That study also showed that the reduction in LH and testosterone as well as in the size of the gonads started by week 4 and it was maximal by week 12. However, at week 20 the parameters were similar to those recorded at week 4, suggesting that the vaccine effects were reverting. However, the length of the study [48] does not confirm this hypothesis or establish the schedule for revaccination.

Figure 5 summarizes the analysis to the application of antiGnRH vaccines as contraceptives in dogs.

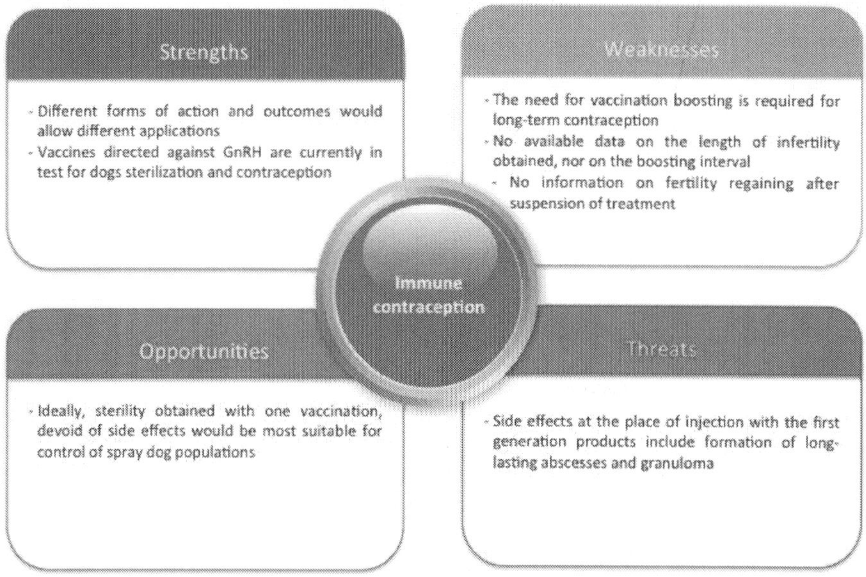

Figure 5. SWOT analysis to the use of antiGnRH vaccines as a contraceptive in dogs.

Immunocontraception holds great promise for canine contraception; still, several drawbacks need to be overcome before being widely introduced into the veterinary practice. The need for regular revaccination may not be an issue in owned dogs but important questions needing answers respect the maintenance of fertility and the time to fertility restoration at the withdrawal of the treatment.

4. Final comments

In this chapter, we have discussed that different methods are available for contraception in domestic dogs. Each one has its own advantages and disadvantages, which should be taken into consideration when discussing with the owner the best therapeutic option available for a particular bitch. When selecting the contraceptive treatment(s) to discuss with the owner, it is important to establish the purpose for the treatment (short- or long-term treatments vs. sterilization), the schedule of drug administration and the costs of the therapeutics, as well as the expected length of treatment and the objectives for the female fertility at the end of the treatment. It is also important to ascertain the dog category (ownerless or community vs. owned dog), the compliance of the owner with the schedule and its expectations toward the meaning of "chemical spaying." And, most of all, it is important to be confident that the selected treatment is adequate to the age of female and neither compromise an existing, undiagnosed, pregnancy, nor trigger or aggravate an existing disease.

Nowadays, safer and healthier alternatives to progestins are already available for medical contraception in dogs that are intended to be breed later in their lives. Major drawbacks for the use of progestins include their deleterious effects on the uterus and mammary gland and the possibility of fertility loss after chronic treatments. Bitches over the age of 5 years are poor candidates for progestin treatments. Good alternatives already exist, and should be recommended whenever the costs of the drug do not impose constraints. Most products can be applied independently of the age of the female and are devoid of side effects on the uterus or the mammary gland.

The identification of new safer methods for nonsurgical contraception in dogs, manufactured at a desirable scale to be provided at affordable rates and needing fewer applications (particularly for permanent and the long-term contraception) is still a challenge to the industry. Active research and increased knowledge in this field promises to change the paradigm of canine contraception in the future.

Acknowledgements

This study was funded by the projects UID/CVT/00772/2013 and UID/CVT/00772/2016 supported by the Portuguese Science and Technology Foundation (FCT).

Author details

Rita Payan-Carreira[1*], Paulo Borges[2] and Alain Fontbonne[2]

*Address all correspondence to: rtpayan@gmail.com

1 CECAV (Animal and Veterinary Sciences Research Centre), Universidade de Trás-os-Montes e Alto Douro, Vila Real, Portugal

2 CERCA, ENVA, Maisons Alfort, France

References

[1] Brown G. Advances in reproductive control technology. In: Association AV, editor. 13th National Urban Animal Management Conference, Caloundra, Queensland; 2003. p. 107–9.

[2] Concannon P. Estrus suppression in the bitch. In: Bonagura JD, Twedt DC, editors. Kirk's Current Veterinary Therapy: XIV: Small Animal Practice. XIV ed. St. Louis, MO: Elsevier Health Sciences; 2009. p. 1024–30.

[3] van Goethem B, Schaefers-Okkens A, Kirpensteijn J. Making a rational choice between ovariectomy and ovariohysterectomy in the dog: a discussion of the benefits of either technique. Vet Surg. 2006;35:136–43.

[4] Kutzler M, Wood A. Non-surgical methods of contraception and sterilization. Theriogenology. 2006;66:514–25.

[5] Purswell BJ. Targets and historical approaches to non-surgical sterilization in dogs and cats. In: ACC&D, editor. ACC&D 4th International Symposium on Non-Surgical Methods of Pet Population Control. Dallas, TX: Alliance for Contraception in Cats & Dogs; 2010.

[6] Concannon PW. Reproductive cycles of the domestic bitch. Anim Reprod Sci. 2011;124:200–10.

[7] Root Kustritz MV. Managing the reproductive cycle in the bitch. Vet Clin North Am Small Anim Pract. 2012;42:423–37,

[8] Jöchle W, Andersen AC. The estrous cycle in the dog: a review. Theriogenology. 1977;7:113–40.

[9] Linde-Forsberg C, Wallén A. Effects of whelping and season of the year on the interoestrous intervals in dogs. J. Small Anim Pract. 1992;33:67–70.

[10] Sokolowski JH. Reproductive patterns in the bitch. Vet Clin North Am. 1977;7:653–66.

[11] Gobello C. Prepubertal and pubertal canine reproductive studies: conflicting aspects. Reprod Domest Anim. 2014;49:e70–3.

[12] Schaefers-Okkens AC, Kooistra HS. Ovaries. In: Rijnberk A, Kooistra HS, editors. Clinical Endocrinology of Dogs and Cats. Hannover, Germany: Schlütersche; 2010. p. 203–34.

[13] Okkens AC, Kooistra HS. Anoestrus in the dog: a fascinating story. Reprod Domest Anim. 2006;41:291–6.

[14] Concannon PW. Endocrinologic control of normal canine ovarian function. Reprod Domest Anim. 2009;44(Suppl 2):3–15.

[15] Massei G, Miller LA. Nonsurgical fertility control for managing free-roaming dog populations: a review of products and criteria for field applications. Theriogenology. 2013;80:829–38.

[16] Johnston. Disorders of the canine uterus and uterine tubes (oviducts). In: Johnston SD, Root Kustritz MV, Olson PS, editors. Canine and Feline Theriogenology. 1st ed. Philadelphia: Saunders; 2001. p. 206–24.

[17] Romagnoli S, Concannon P. Clinical use of progestins in bitches and queens: a review. In: Concannon PWE, England G, Verstegen J III, Linde-Forsberg, C, editor. Recent

Advances in Small Animal Reproduction. Ithaca, New York: International Veterinary Information Service (http://www.ivis.org); 2003.

[18] Romagnoli S, Sontas BH. Prevention of breeding in the female. In: England GCW, von Heimendahl A, editors. Manual of Small Animal Reproduction and Neonatology. 2nd ed. Gloucester, UK: BSAVA Publishing; 2010. p. 23–33.

[19] Max A, Jurka P, Dobrzynski A, Rijsselaere T. Non-surgical contraception in female dogs and cats. Acta Sci. Pol. Zootech. 2014;13:3–14.

[20] England G. Pharmacological control of reproduction in the dog and bitch. In: Simpson G, England G, Harvey J, editors. BSAVA Manual of Small Animal Reproduction and Neonatology. Birmingham: British Small Animal Veterinary; 1998. p. 197–218.

[21] Sekeles E, de Lange A, Samuel L, Aharon DC. Oestrus control in bitches with chlormadinone acetate. J. Small Anim. Pract. 1982;23:151–8.

[22] Phillips TC, Larsen RE, Hernandez J, Strachan L, Samuelson D, Shille VM, et al. Selective control of the estrous cycle of the dog through suppression of estrus and reduction of the length of anestrus. Theriogenology. 2003;59:1441–8.

[23] Trigg TE, Wright PJ, Armour AF, Williamson PE, Junaidi A, Martin GB, et al. Use of a GnRH analogue implant to produce reversible long-term suppression of reproductive function in male and female domestic dogs. J Reprod Fertil Suppl. 2001;57:255–61.

[24] Borges P, Fontaine E, Maenhoudt C, Payan-Carreira R, Santos N, Leblond E, et al. Fertility in adult bitches previously treated with a 4.7 mg subcutaneous deslorelin implant. Reprod Domest Anim. 2015;50:965–71.

[25] Navarro C, Schober P. Pharmacodynamics and pharmacokinetics of a sustained-release implant of deslorelin in companion animals. In: England G, Kutzler M, Comizzoli P, Nizanski W, Rijsselaere T, Concannon P, editors.7th International Symposium on Canine and Feline Reproduction. Whistler, BC, Canada: ISCFR; 2012. p. 177–8.

[26] Gobello C. New GnRH analogs in canine reproduction. Anim Reprod Sci. 2007;100:1–13.

[27] Maenhoudt C, Santos NR, Fontbonne A. Suppression of fertility in adult dogs. Reprod Domest Anim. 2014;49(Suppl 2):58–63.

[28] van Haaften B, Bevers MM, van den Brom WE, Okkens AC, van Sluijs FJ, Willemse AH, et al. Increasing sensitivity of the pituitary to GnRH from early to late anoestrus in the beagle bitch. J Reprod Fertil. 1994;101:221–5.

[29] Wolf T, Meyer H, Kutzler M. Litter size response to oestrous induction with deslorelin (Ovuplant®) in dogs. Reprod Domest Anim. 2012;47(Suppl 6):387–8.

[30] Wright PJ, Verstegen JP, Onclin K, Jöchle W, Armour AF, Martin GB, et al. Suppression of the oestrous responses of bitches to the GnRH analogue deslorelin by progestin. J Reprod Fertil Suppl. 2001;57:263–8.

[31] Corrada Y, Hermo G, Johnson CA, Trigg TE, Gobello C. Short-term progestin treatments prevent estrous induction by a GnRH agonist implant in anestrous bitches. Theriogenology. 2006;65:366–73.

[32] Sung M, Armour AF, Wright PJ. The influence of exogenous progestin on the occurrence of proestrous or estrous signs, plasma concentrations of luteinizing hormone and estradiol in deslorelin (GnRH agonist) treated anestrous bitches. Theriogenology. 2006;66:1513–7.

[33] Fontaine E, Mir F, Vannier F, Gérardin A, Albouy M, Navarro C, et al. Induction of fertile oestrus in the bitch using Deslorelin, a GnRH agonist. Theriogenology. 2011;76:1561–6.

[34] Volkmann DH, Kutzler MA, Wheeler R, Krekeler N. The use of deslorelin implants for the synchronization of estrous in diestrous bitches. Theriogenology. 2006;66:1497–501.

[35] Arlt SP, Spankowsky S, Heuwieser W. Follicular cysts and prolonged oestrus in a female dog after administration of a deslorelin implant. N Z Vet J. 2011;59:87–91.

[36] Palm J, Reichler IM. The use of deslorelin acetate (Suprelorin®) in companion animal medicine. Schweiz Arch Tierheilkd. 2012;154:7–12.

[37] Ackermann CL, Volpato R, Destro FC, Trevisol E, Sousa NR, Guaitolini CR, et al. Ovarian activity reversibility after the use of deslorelin acetate as a short-term contraceptive in domestic queens. Theriogenology. 2012;78:817–22.

[38] Maenhoudt C, Santos NR, Fontaine E, Mir F, Reynaud K, Navarro C, et al. Results of GnRH agonist implants in oestrous induction and oestrous suppression in bitches and queens. Reprod Domest Anim. 2012;47(Suppl 6):393–7.

[39] Rubion S, Desmoulins PO, Rivière-Godet E, Kinziger M, Salavert F, Rutten F, et al. Treatment with a subcutaneous GnRH agonist containing controlled release device reversibly prevents puberty in bitches. Theriogenology. 2006;66:1651–4.

[40] Schäfer-Somi S, Kaya D, Gültiken N, Aslan S. Suppression of fertility in pre-pubertal dogs and cats. Reprod Domest Anim. 2014;49(Suppl 2):21–7.

[41] Trigg TE, Doyle AG, Walsh JD, Swangchan-uthai T. A review of advances in the use of the GnRH agonist deslorelin in control of reproduction. Theriogenology. 2006;66:1507–12.

[42] Goericke-Pesch S, Wehrend A, Georgiev P. Suppression of fertility in adult cats. Reprod Domest Anim. 2014;49(Suppl 2):33–40.

[43] Valiente C, Romero GG, Corrada Y, de la Sota PE, Hermo G, Gobello C. Interruption of the canine estrous cycle with a low and a high dose of the GnRH antagonist, acyline. Theriogenology. 2009;71:408–11.

[44] Purswell BJ, Kolster KA. Immunocontraception in companion animals. Theriogenology. 2006;66:510–3.

[45] Walker J, Ghosh S, Pagnon J, Colantoni C, Newbold A, Zeng W, et al. Totally synthetic peptide-based immunocontraceptive vaccines show activity in dogs of different breeds. Vaccine. 2007;25:7111–9.

[46] Fagerstone KA. Mechanism of GnRH contraceptive vaccine-mediated infertility and its applications In: ACC&D, editor. ACC&D 3th International Symposium on Non-Surgical Methods of Pet Population Control. Alexandria, Virginia; 2006. pp. 4.

[47] Vargas-Pino F, Gutiérrez-Cedillo V, Canales-Vargas EJ, Gress-Ortega LR, Miller LA, Rupprecht CE, et al. Concomitant administration of GonaCon™ and rabies vaccine in female dogs (Canis familiaris) in Mexico. Vaccine. 2013;31:4442–7.

[48] Donovan CE, Greer M, Kutzler MA. Physiologic responses following gonadotropin-releasing hormone immunization in intact male dogs. Reprod Domest Anim. 2012;47(Suppl 6):403–5.

Section 5

Orthopedics

Chapter 8

Patellar Luxation in Small Animals

Cleuza M.F. Rezende, Renato César Sachetto Tôrres,
Anelise Carvalho Nepomuceno,
Juliana Soares Lara and
Jessica Alejandra Castro Varón

Additional information is available at the end of the chapter

http://dx.doi.org/10.5772/65764

Abstract

This study describes lesions that occur in the stifle joints of dogs with patellar luxation. These lesions are associated with the animal's age, body weight and degree of luxation. It also reports on the rate of re-dislocation. The patellar lesions found include articular cartilage erosion, subchondral bone exposure, a flattened or concave patellar surface and enthesophytes. Extrapatellar lesions included synovitis, osteophytes, blunting of the trochlear groove, an absent trochlea, erosion of the condylar margins, capsule thickening, a long digital extensor tendon injury, "joint mice," flap formation, cranial cruciate ligament rupture and meniscal prolapse. Such lesions were frequently found in animals with grade II or III luxation that were aged 24 months or older; they were more severe in dogs weighing more than 15 kg. Patellar luxation causes changes that favour articular degeneration and should be treated surgically. Conservative treatment relieves pain but does not address tissue alterations.

Keywords: luxation, patella, dogs, lesions, joint

1. Introduction

Patellar luxation is a frequent occurrence in dogs and thus represents a common finding in everyday veterinary trauma and orthopaedic practice [1]. However, early treatment is not considered important. Clinical signs vary with the severity of luxation, and in some cases, the diagnosis is made during routine physical examination. The luxation may be present at birth, and in these cases are usually grades III and IV associated with severe skeletal deformities with functional disability (**Figure 1A, B**). In puppies, the surgical correction must be performed between 1 and 3 months of age, and not later than 3 months [2]. At this age, the skeletal deformities can be reversed after alignment of the limb. Patellar luxation may occur at birth, during growth or at a later stage. The pathophysiology of congenital luxation remains a topic of discussion [3–5]; a consequence of complex skeletal abnormalities that alter the limb alignment is considered [5]. The condition may be unilateral or bilateral and can be asymptomatic, and most cases are medial [3, 5]. Occasionally, the luxation can occur in both directions in the same joint [3], which is a surgical challenge. It is possible to find medial patellar luxation in one joint and lateral luxation in another joint in the same dog. Some cases exhibit patellar subluxation. Although much less frequently than in dogs, patellar dislocation also occurs in cats [3].

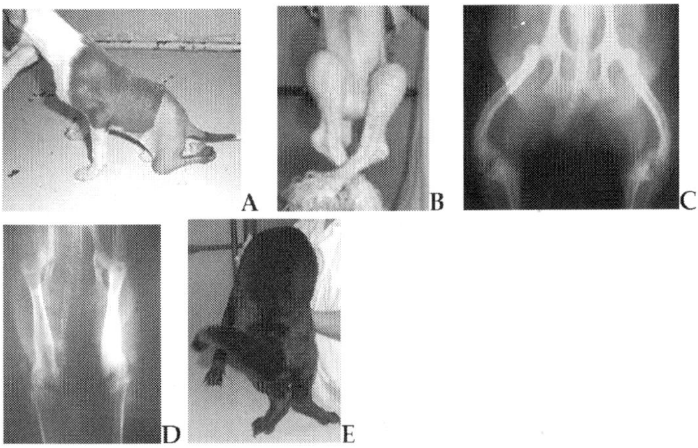

Figure 1. (A) Abnormal position associated with bilateral grade IV medial patellar luxation in puppies 60 and (B) 90 days old; (C) ventrodorsal radiograph, showing bilateral medial luxation of the patella, and angular deformities in a 90-day-old poodle; (D) ventrodorsal radiograph, showing bilateral lateral luxation of the patella in a 9-month-old dog breed Sharpei; (E) abnormal stance associated with bilateral grade IV lateral patellar.

The incidence of severe articular lesions found during routine surgeries in small-, medium- or large-breed dogs presenting with patellar luxation is high. In some cases, the patella injury is so serious that correction is not possible or prudent. In these instances, a better treatment option may be prosthetic replacement. Patellectomy is suggested as a treatment alternative in some severe cases [5, 6], but the removal of the patella does not correct the skeletal changes [6], calling this therapeutic option into question.

According to the literature [5], surgery might not be necessary in cases without clinical manifestations or when lameness is mild; however, even under such circumstances, the joint damage is irreversible.

Patellar luxation is a degenerative condition, and surgical treatment should be performed as early as possible, while clinical signs are mild or even before they appear. The aims of this

http://dx.doi.org/10.5772/65764

study were to perform a retrospective survey of the lesions found in the stifle joints of dogs with patellar luxation, to investigate the associations between these lesions and the animal's age, body weight and degree of luxation and to estimate the incidence of re-luxation after surgical treatment.

2. Anatomy and etiopathogenesis

The patella is an oval-shaped sesamoid bone that is located in the quadriceps tendon [7, 8] and connected to the fabella by a thin band of loose connective tissue known as the medial and lateral patellofemoral ligament [3, 7, 8]. The tendon located between the distal aspect of the patella and the tuberosity of the tibia is known as the patellar ligament [3]. The patella acts as a lever arm favouring the extension of the quadriceps [3], and the patellofemoral joint increases the mechanical efficiency of the quadriceps mechanism [8]. The correct alignment of the quadriceps femoris, patella, trochlea, patellar ligament and tibial tuberosity prevents patellar luxation or subluxation [8]. An adequate supply of articular cartilage depends on a normal joint between femoral trochlea and patella. Patellar luxation causes articular cartilage degeneration [3].

The pathogenesis of developmental patellar luxation remains speculative and controversial [3, 5, 9]. Based on the thesis by Putnam [10], abnormal femoral and neck angles of inclination and anteversion are proposed to influence development and eventual malformation of the pelvic limb. However, this has not been proven [11].

The III and IV degrees of luxation are associated with severe angular and rotational deformities of the femur and tibia [8], which are more pronounced in medial luxation; the limb has an S-shaped conformation (**Figure 1B, C**). Some cases are difficult to treat, with poorer prognosis. Corrective osteotomy must be considered in these cases to restore the normal alignment of the quadriceps complex [5, 6]. It is sometimes necessary to also remove up to 1.0 cm of bone extension to return the patella to the femoral trochlea. In these cases, patella tibial tuberosity alignment is achieved but paw rotation remains. Therefore, early intervention should be considered in cases of patellar luxation independent of the existence of clinical signs.

Clinical evidence shows that intermittent patellar luxation, contrary to the logical assumption, produces varying degrees of injury to the articular surfaces of the patella and the femoral condyle, causing degeneration of articular cartilage. This may also result in a flattening of the femoral condyle and consequently facilitate dislocation [8]. The position of the patella relative to the femoral trochlea, more proximal, is considered by some authors [8, 13, 14] to be an important factor in canine patellar luxation.

Although seldom discussed in the literature, subluxation also occurs and is observed in clinical practice. The affected dogs are typically adults presenting with pain and lameness due to the wear on the patella [8]. This condition is difficult to diagnose, especially in heavy English Bulldogs; a careful physical examination is necessary. Other findings, including cruciate ligament rupture and even the presence of intra-articular loose bodies and flap formation, can be seen in cases of chronic patellar dislocation.

Medial patellar dislocation is responsible for approximately 80% [5] of cases, with bilateral involvement in up to 50–65% [5, 14]. Lameness is one of the presenting clinical signs; however, it is subjective; lameness is seen in other conditions that affect the hind limb, which should be included in the differential diagnosis [6, 15]. Medial luxation is considered characteristic of small breeds, but it has been reported with increasing frequency in large and giant breeds [5].

Although many dogs present at 6–12 months of age, some animals with mild lameness present later, with rupture of the cranial cruciate ligament or pain caused by osteoarthritis [8]. Often, the patellar luxation is not considered to be an important joint disorder until it becomes severe. In puppies, the situation is even more critical. Post-operative studies show that it is possible to correct the limb deformities and achieve complete reversal of skeletal changes if the surgery is performed before 60 days of age [2].

Lateral luxation also occurs in small dogs; the skeletal deformities are the reverse of those seen with medial luxation (**Figure 1D, E**), demonstrating an increased angle of anteversion, coxa valga, medial torsion of the distal femur, lateral condylar dysplasia, lateral rotation of the tibia and external rotation of the paw [16, 17]. The articular cartilage injuries in both types of luxation result in osteoarthritis [16, 18], which is usually mild-to-moderate and unrelated to the degree of luxation or lameness [19]. Lateral luxation of the patella may cause injury at the origin of the long digital extensor tendon, leading to rupture of the tendon in severe cases.

3. Clinical signs

The clinical signs of patellar luxation vary according to the degree of deformity, duration of the condition, unilateral or bilateral stifle involvement [8] and the age of onset. Degree III or IV luxation in puppies prevents ambulation (**Figure 1A, B**) and causes the characteristic changes of Genu varum or a "bow-legged" conformation in some cases of medial patellar luxation (**Figure 2A, B**). In lateral patellar luxation, the associated abnormal anatomic features are reversed. These dogs may have a knock-kneed appearance when affected bilaterally (**Figure 2C, D**).

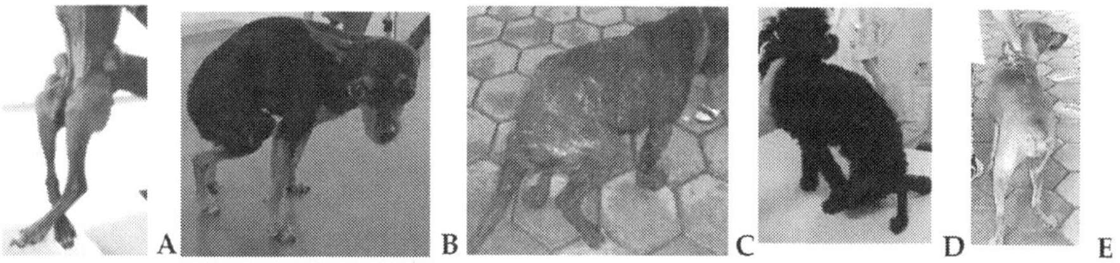

Figure 2. Abnormal stance associated with: bilateral grade IV medial patellar luxation in a (A) 5- and (B) 3-year-old pinscher; lateral patellar luxation in a 6-month-old Fila Brasileiro (C) and (D) 3-year-old poodle; (E) unilateral grade III lateral luxation in a crossbreed 4-year-old dog.

http://dx.doi.org/10.5772/65764

In clinical practice, four classes of patients [5] are encountered: neonates and older puppies with grade III or IV luxation and inability to walk (**Figure 1A**), young or mature dogs with grade II to III luxation and mild clinical signs for a long period of time until presenting when the clinical condition worsens, older animals with grade I or II luxation and sudden claudication due to the cruciate cranial ligament rupture or degenerative joint disease pain and asymptomatic dogs.

Clinical signs vary from animal to animal and may be intermittent or continuous. Associated joint damage and overweight may worsen the clinical signs. The physical examination should be performed carefully; gentle palpation without causing pain is the goal, considering the difficulty of identifying the patella in toy and miniature animals as well as in dogs with severe deformities. In general, dogs with lateral luxation have more ambulation problems than those with medial luxations [5].

The physical examination should consider aspects such as instability in both directions, crepitus, degree of tibial rotation, limb deformity, inability to reduce the patella, location of the reduced patella within the trochlea, inability to stand the limb to a normal standing angle and presence/absence of cruciate ligament rupture. This information is necessary for surgical planning [5].

Patellar dislocation is classified into four grades [20] to facilitate the diagnosis and plan the method of surgical repair:

Grade I. A dog with grade I patellar luxation rarely shows lameness and carries the limb occasionally. The patella can be manually luxated when the stifle is extended, but it returns to the trochlea when released. No crepitation is apparent. Internal rotation of the tibia and displacement of tibial tuberosity are minimal.

Grade II. Luxation occurs more frequently than in grade I. Lameness signs are usually intermittent and mild. The patella moves easily, especially when the foot is rotated, while the patella is pushed. The proximal tibial tuberosity may be rotated up to 30° with medial luxations and less with lateral luxations. Many grade II patients live reasonably well for many years, but the injuries from constant friction between the patella and femoral condyle result in crepitation and increasing discomfort [5].

Grade III. The patella is permanently luxated, with torsion of the tibia and deviation of the tibial crest between 30° and 60°. It can be reduced, but luxation recurs immediately. The trochlea is very shallow or even flattened. Although they have permanent luxation, many animals use the limb with the stifle held in a semi-flexed position.

Grade IV. The patella is permanently luxated, and it is not possible to manually reposition it in the trochlea. It lies just above the medial condyle (if a medial luxation) (**Figure 1C**). Angular and rotational deformity of the femur and tibia are generally marked, and the tibial crest is displaced 60°–90°. The trochlea is shallow, absent or even convex. The limb may be carried if unilateral, or the animal moves in a crouched position, with the limbs partially flexed or carried in toy animals (**Figure 2B**).

4. Radiography

The diagnosis of patellar luxation is clinical, but radiography may help to confirm the diagnosis, showing the luxated patella in more severe cases, and can demonstrate any bony deformities that are present [6]. Mediolateral and craniocaudal radiographs enable assessment of femoral and tibial deformities (**Figure 3A, B**). Tangential views of the flexed stifle enable assessment of the femoral trochlea and its depth (**Figure 3C**). Radiographs can also delineate the morphological changes of the patella and trochlea, demonstrate secondary osteoarthritis [6, 8] and allow the prognosis of limb function.

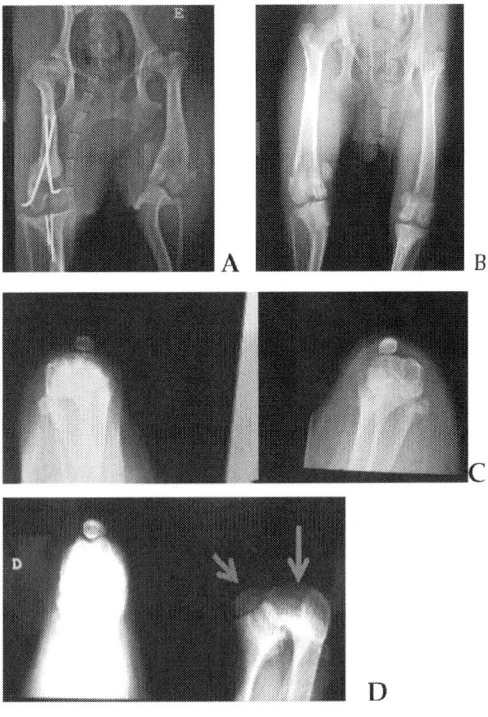

Figure 3. Anteroposterior radiograph of the limbs of a dog, showing: (A) medial luxation grade IV in a 16-month-old Pekinese (left); (B) bilateral medial luxation grade III; (C) tangential view of flexed stifle showing an absent trochlea; (D) tangential view showing shallow trochlea (blue arrow), and ectopic patella in a new trochlea (red arrow).

5. Treatment

Surgical management of patellar luxation in asymptomatic small-breed dogs is controversial and is not considered necessary. However, surgery is recommended early in young puppies

(3–4 months) prior to irreparable contracture and in medium to large breeds prior to erosion and trochlea deformity [5]. The goal of surgical treatment is to realign the extensor apparatus to restore normal stifle biomechanics and stabilize the femoropatellar joint. Surgical procedures to stabilize patellar luxation can be divided into soft tissue reconstruction and bone reconstruction [5]. The surgical techniques include deepening of the femoral trochlea, tibial tuberosity transposition, medial soft tissue release and lateral soft tissue tightening. Other procedures, such as transplantation of the origin of the rectus femoris muscle and corrective osteotomy, can be necessary in grade III or IV patellar luxation [6, 8]. It is sometimes necessary to shorten the femur to reduce the luxation. The presence of patellar injuries does not necessarily indicate a surgical contraindication. Healing may occur, and the realignment of the quadriceps shows favourable results even in the presence of osteoarthritis. Patellectomy is rarely necessary, does not produce benefits and does not correct the alignment of the quadriceps complex [6].

6. Patellar luxations in cats

A slight increase in the occurrence of patellar luxation has been observed in cats; however, it is far less common than in dogs [21]; it has been reported in breeds such as the Abyssinian, Devon Rex, Siamese, domestic shorthair cats [22–24], and non-pedigree cats [21, 25]. Some authors report a relationship with hip dysplasia [8], but this remains controversial. As in the dog, medial patellar luxation is the most common and may be either unilateral or bilateral. Most affected cats are usually relatively young at the time of presentation [25]. Some anatomical considerations are necessary to prevent misdiagnosis. The patella in the cat is relatively wider and has more physiological laxity than in dogs. It can be manually moved onto the trochlear ridge of the femoral condyles in many normal cats. Therefore, the grading system developed for dogs should be used with caution [26].

Considering these peculiarities, Voss et al. [27] suggested the following grading system for patellar luxation in cats:

Grade A. Patella can be completely luxated with digital pressure, but immediately returns into position after pressure is released.

Grade B. Patella can be completely luxated with digital pressure and remains temporarily luxated after the pressure is released.

Grade C. Patella luxates when the tibia is internally rotated without exerting direct digital pressure.

Grade D. Patella is temporarily or permanently luxated without any manipulation.

Anatomical changes are similar to those in the dog, including a shallow trochlear groove and medial displacement of the tibial tuberosity. Severe conformational changes are rarely observed, and secondary osteoarthritis is generally absent or mild [28]. It may be associated with the age at diagnosis of patellar luxation because older cats may have severe degenerative changes in their stifle joints [28]. Erosions in the patella and on the condylar ridge may be seen as in dogs. Although surgical treatment is suggested when the luxation results in lameness [21], it should be considered early, aiming to align the quadriceps and prevent degenerative changes.

Congenital bilateral patellar aplasia has been described in two Siamese cat littermates [29].

7. Materials and methods

This was a retrospective study that assessed the clinical surgical records of dogs treated for patellar luxation from January 2005 to January 2016 at the Veterinary Hospital, Federal University of Minas Gerais (FUMG) in Brazil. Data describing age, body weight and the degree of patellar luxation at admission were collected for each animal. The animals were then categorized based on body weight (≤5, 5–15 and ≥15 kg), age (<12, 12–24 and >24 months) and degree of luxation according to Putnam's [10] classification as adapted by Singleton [19]. In addition, data relative to the intraoperative period were recorded; these included the presence of stifle joint damage, classified as patellar or extrapatellar lesions and the frequency of post-operative re-luxation. The associations between patellar or extra-patellar lesions and the animal's age, body weight and degree of luxation were investigated. Patellar lesions included the presence of cartilage erosion, the extent of the erosion (one-fourth, one-half or the full patellar articular surface), the morphology of the patella (concave or flat), the exposure of subchondral bone and the presence of enthesophytes. Extra-patellar lesions included the presence of erosion and subchondral bone exposure in the medial or lateral femoral condyles, the presence of osteophytes, synovitis, capsular thickening or shallowing of the trochlea, an absent or convex trochlear groove, cruciate ligament rupture and injury of the long digital extensor tendon and the menisci.

The frequency of lesions in the stifle joint was subjected to descriptive analysis, and the rate of patellar re-luxation was assessed using the chi-square test. The significance level was set at $P < 0.05$.

This retrospective study also included the clinical records of eight cats (13 joints) treated for patellar luxation. Data describing age, body weight and the degree of patellar luxation at admission were collected for each animal.

8. Results

A total of 280 luxated joints from 202 dogs were assessed; 244 were in medial luxation, whereas 36 were in lateral luxation. Sixteen (5.7%) were classified as grade I, 118 (42.1%) as grade II, 57 (20.4%) as grade III and 89 (31.8%) as grade IV.

The patellar lesions identified included cartilage erosion of one-fourth (15.0%), one-half (11.8%) or all (1.8%) of the patella, exposure of subchondral bone (4.3%), a flattened or concave patellar surface (15.0%) (**Figure 4**), the presence of enthesophytes (12.1%), cartilage flap at the edge of femoral condyle, joint mice and lesions caused by the patella rubbing on the long digital extensor tendon (**Figure 5**). **Tables 1** and **2** show the lesions according to degree of dislocation, weight and age.

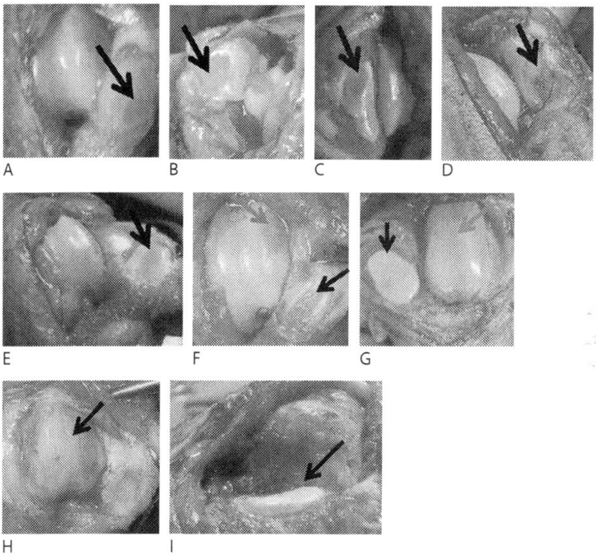

Figure 4. Photograph of the stifle joint of a dog subjected to surgery for patellar luxation. Notice the erosion of the patellar articular surface (black arrow). (A) Superficial erosion; (B) moderate erosion, (C and D) severe erosion; (E) severe erosion in the whole extension of the patella; (F) erosion of the edge of the medial condyle and erosion of the patella with subchondral bone exposure (blue and black arrows); (G) notice the flattened patellar surface (black arrow) and shallow trochlear groove (blue arrow); (H) convex femoral trochlea (black arrow); (I) patella concave.

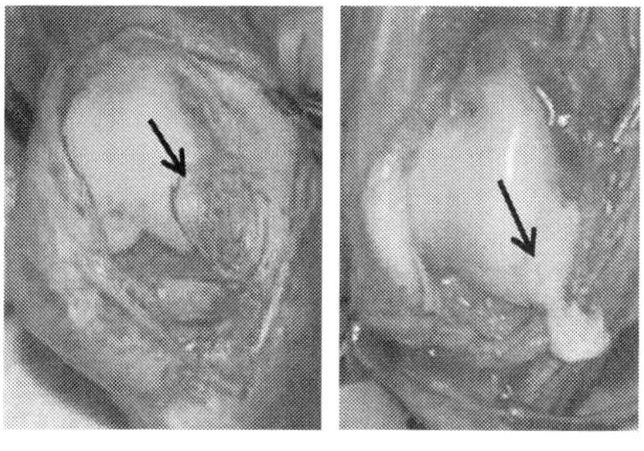

Figure 5. Photograph of the stifle joint of a dog subjected to surgery for patellar luxation. Notice the lesion by rubbing of the patella on the long digital extensor tendon (black arrow).

One rare instance of resorption of the patella was diagnosed in a dog with a grade III luxation. Some remnants of the bone were found adhered to the patellar tendon, while the central area was soft and exhibited loss of bone and cartilage. A 1-year history of severe lameness with frequent non-weight-bearing and grade II medial luxation in the contralateral limb existed. Resorption of the patella was treated with implantation of polyhydroxybutyrate patellar prosthesis. The outcome was favourable.

	Cartilage erosion												
	¼ patella		**½ patella**		**Full patella**		**Subchondral bone exposure**		**Flattened/concave patella**		**Enthesophytes**		**Assessed joints**
Severity	**n°**	**%**	**n°**	**%**	**n°**	**%**	**n°**	**%**	**n°**	**%**	**n°**	**%**	**n°**
Grade I	1	6.3	0	0	0	0	0	0	0	0	0	0	16
Grade II	22	18.6	17	14.4	0	0	6	5.1	12	10.1	11	9.3	118
Grade III	13	22.8	10	17.5	2	4.1	5	8.8	11	19.3	15	26.3	57
Grade IV	6	6.7	6	6.7	3	3.4	1	1.1	19	21.3	8	9.0	89
Total	42		33		5		12		42		34		280
Weight													Assessed dogs (n°)
<5 kg	22	21.4	5	6.0	3	2.8	0	0	14	13.0	13	12.0	108
5–15 kg	11	21.2	12	23.1	1	1.9	5	9.6	9	17.3	10	19.3	52
>15 kg	6	20.0	10	33.3	2	6.7	5	16.7	7	23.3	6	20.0	30
Total	39		27		6		10		30		29		190*
Age													(n°)
<12 months	6	7.0	5	5.8	3	3.5	5	5.8	13	15.1	5	5.8	86
12–24 months	3	15.0	1	5.0	0	0	0	0	1	5.0	3	15	20
>24 months	27	30.0	22	24.4	4	4.4	7	7.8	16	17.8	20	22.2	90
Total	36		28		7		12		30		28		196**

* The weight was not reported in 12 animals that were excluded from this assessment.
** The age was not reported in six animals that were excluded from this assessment.

Table 1. Estimated patellar lesions in dogs with patellar luxation treated at the Veterinary Hospital, FUMG, 2005–2016 according to the degree of luxation, body weight and age.

The frequency of cartilage erosion affecting one-fourth or one-half of the patella accompanied by subchondral bone exposure was higher among those with grade II or III luxation. Erosion of the full patellar surface occurred only in grade III or IV luxation. Patellar lesions were frequently observed on the lateral surface in medial luxation, but were seen on the medial surface in lateral luxation. Anatomical changes of the patella, such as a flattened or concave surface, occurred primarily in grade III or IV luxation, whereas enthesophytes were

http://dx.doi.org/10.5772/65764

most frequently found in grade III luxation. In animals older than 24 months, the lesions most frequently found were cartilage erosion affecting one-fourth or one-half of the patella, subchondral bone exposure and the presence of enthesophytes. The frequency of anatomical changes of the patella (flattened or concave surface) was highest among those older than 24 months of age and lowest in those aged 12–24 months. The frequency of patellar lesions was proportionally higher among those animals weighing more than 15 kg (**Table 1**). Joint mice and flap formation (**Figure 6**) were observed in two dogs with patellar luxation grade III. One dog was 6 months of age and weighed 22 kg and the other was 10 years old and weighed 5.8 kg.

Extra-patellar articular lesions	**Synovitis**		**Osteophytes**		**Shallow trochlea**		**Absent trochlea**		**Femoral condyle erosion**		**Thickened capsule**		**Subchondral bone exposure**		**Assessed joints**
Severity	**n°**	**%**	**n°**	**%**	**n°**	**%**	**n°**	**%**	**n°**	**%**	**n°**	**%**	**n°**	**%**	**n°**
Grade I	6	37.5	1	6.25	0	0	0	0	0	0	0	0	0	0	16
Grade II	41	34.7	37	31.4	56	47.5	4	3.7	41	34.7	25	21.2	4	3.7	118
Grade III	18	31.6	20	35.1	27	47.4	5	8.8	34	59.6	21	36.8	8	14.0	57
Grade IV	17	19.1	15	16.9	40	44.9	27	30.3	25	28.1	20	22.5	10	11.2	89
Total	82		73		123		36		100		66		22		280
Weight															Assessed dogs (n°)
<5 kg	20	18.5	20	18.5	42	38.9	17	15.7	25	23.1	17	15.7	3	2.8	108
5–15 kg	21	40.4	17	32.7	29	55.8	4	7.7	24	46.2	16	30.8	5	9.6	52
>15 kg	12	40	17	56.7	15	50.0	1	3.3	29	96.7	16	53.3	8	26.7	30
Total	53		54		86		22		78		49		16		190
Age															n°
<12 months	10	11.6	6	7.0	43	50.0	18	20.9	16	18.6	14	16.3	8	9.3	86
12–24 months	7	35.0	3	15.0	6	30.0	1	5.0	5	25.0	4	20.0	0	0	20
>24 months	33	36.7	43	47.8	40	44.4	4	4.4	36	40.0	23	25.6	9	10.0	90
Total	50		52		89		23		57		41		17		196

Table 2. Estimated extra-patellar articular lesions in dogs with patellar luxation treated at the Veterinary Hospital, UFMG, 2005–2016 according to degree of luxation, body weight and age.

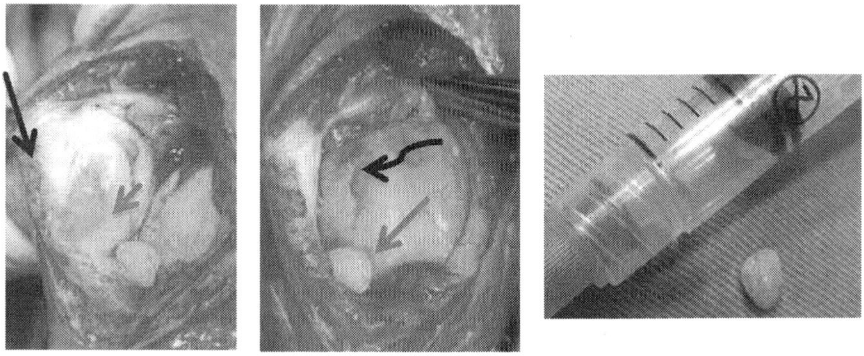

Figure 6. Photography of the stifle joint of a dog to surgery for patellar luxation. Notice the erosion of the patellar articular surface (black arrow), joint mice (blue arrow) and flap formation on the edge of the medial condyle (black S arrow).

Several extra-patellar lesions involving soft or hard tissue were observed. On intraoperative assessment, 29.3% of the joints exhibited synovitis, 26.1% had periarticular osteophytes, 43.9% had shallow trochlear grooves, 12.9% had absent trochlea, 35.7% exhibited cartilage erosion in condylar margins, 7.9% exhibited subchondral bone exposure in condylar margins, 23.6% had thickened capsules (**Table 2**), 3.6% had long digital extensor tendon injuries associated with lateral luxation, 0.7% exhibited joint mice and flap formation, 9.3% had ruptured cranial cruciate ligaments and 3.1% exhibited prolapse of the menisci. Synovitis occurred in 37.5% of the animals with grade I luxation. Periarticular osteophytes and shallow trochlear grooves were most frequently found among the animals with grade II luxation. Erosion of the femoral condylar margins and thickened capsules occurred most frequently among those with grade III luxation, while absent trochlear grooves were most frequently among those with grade IV luxation. Exposure of subchondral bone on the condylar margins predominated among those with grade III or IV luxation. Lesions in medial condyles were observed in joints in medial luxation, whereas lesions in lateral condyles were frequent in joints in lateral luxation. Extra-patellar lesions were proportionally higher among those weighing more than 15 kg.

The frequency of extra-patellar articular lesions according to age, body weight and degree of luxation is presented in **Table 2**.

Patellar re-luxation after surgical repair occurred in 12.9% of the joints distributed across all grades, ranging between 6% and 15%. Statistical analysis using the chi-square test suggests that no statistically significant differences exist between the variable "re-luxation patellar" and "degree of luxation," and "type of luxation (medial or lateral)" ($P > 0.05$).

The frequency of re-luxation according to type and degree of luxation is shown in **Table 3**.

Severity	n	Re-luxation (%)	NO re-luxation (%)
Grade I	16	1 (6.2)[a]	15 (93.8)[a]
Grade II	118	15 (12.7)[a]	103 (87.3)[a]
Grade III	57	7 (12.3)[a]	50 (87.7)[a]
Grade IV	89	13 (14.6)[a]	76(85.4)[a]
Total	280	36 (12.9)	244 (87.1)
Medial luxation	244	32 (13.1)[a]	212 (86.9)[a]
Lateral luxation	36	4 (11.1)[a]	32 (88.9)[a]

[a] In the columns, frequencies with different letters differ between groups using the chi-square test ($P < 0.05$).

Table 3. Estimated patellar re-luxation in dogs treated at the Veterinary Hospital, FUMG, 2005–2016 according to the type (medial or lateral) of luxation and degree of luxation. The rate of patellar re-luxation was assessed using the chi-square test.

A total of 13 luxated joints from eight cats were assessed, and all were medial luxations. Four (30.7%) were classified as grade IV, seven (53.4%) as grade III and two (15.4%) as grade II. Six cats were younger than 12 months and two were 12–24 months of age. Hip dysplasia was present in two cats (25%). Two cats from the same litter had patellar luxation grade IV associated with bone and tail deformity (**Figure 7**). One of them had unilateral patellar aplasia. These cats showed gait abnormalities, but were able to move around the house. Six were mixed-breed cats, one was Siamese, and one was an exotic cat. The surgical treatment was successful, and there was no recurrence of luxation.

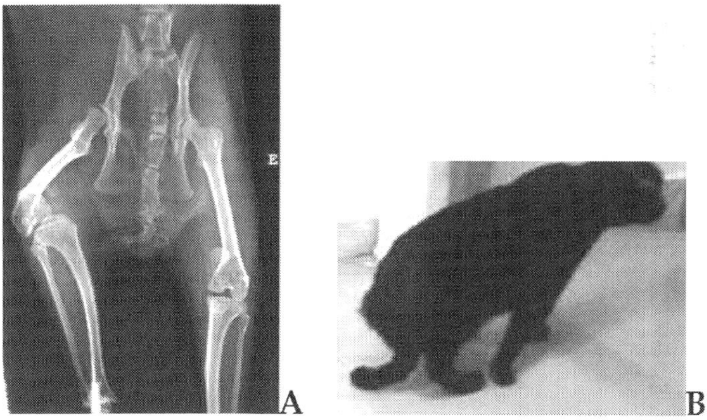

Figure 7. Anteroposterior radiograph of the limbs of a cat, showing: (A) bone deformity, and tail, dysplasia and medial patellar luxation grade III on the left side in a 7-month-old cat; (B) abnormal stance associated with bone deformities and medial patellar luxation grade III.

Lesions such as flattened patellar surface and shallow trochlear groove were also observed in cats with patellar luxation grades III and IV (**Figure 8**).

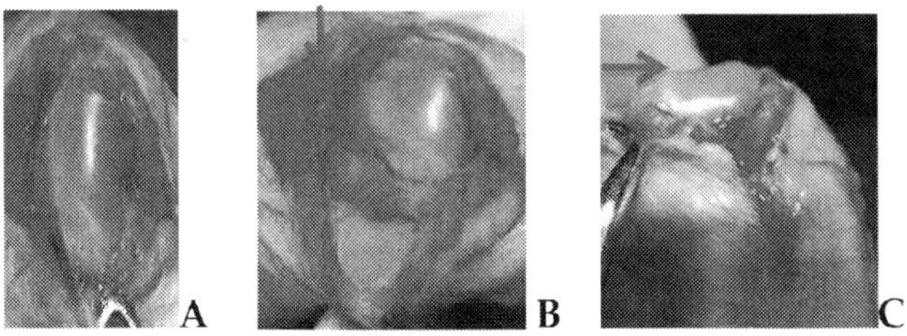

Figure 8. Photography of the stifle joint of a cat to surgery for patellar luxation. Notice the erosion of the edge of the medial condyle (A, blue arrow), flattened patellar surface (B, blue arrow) and shallow trochlear groove (C, blue arrow).

9. Discussion

Patella luxation in small animals is a condition in which early treatment has not been highlighted. The suggested course has been to wait for the manifestation of clinical signs [5, 6, 8, 21], which results in lameness. However, dislocation and subluxation are already characteristic clinical signs of anatomical abnormalities, which certainly will cause articular changes. Subluxation [6] is a clinical challenge, especially in obese and heavy dogs like English Bulldogs, making it difficult to diagnose the cause of progressive lameness [6].

In the literature, there are few descriptions of lesions occurring in the stifle joints of dogs with patellar luxation. Cartilage erosion on the patellar articular surface was reported by Remedios et al. [30]; however, they did not provide information about the extent of the erosion or the weight, age or degree of luxation. Daems et al. [31] reported cartilage erosions mainly in heavier dogs and with grade IV patellar luxation. Destruction of cartilage on the articular surface of the patella in both medial and lateral luxation and marked synovial reaction at the origin of the long digital extensor tendon in lateral luxation were reported by L'Eplattenier and Montavon [3]; Pérez et al. [14] reported erosion in the medial trochlear ridge and shallow trochlear groove.

The results of this study indicate a high frequency of patellar and extra-patellar lesions, especially in grade II and III luxations. According to the literature, although these animals exhibit intermittent and persistent lameness, respectively [1, 5, 32], they continue to use the affected limb for ambulation. These facts account for the larger number of lesions found among animals with grade II or III luxation. The use of the affected limb leads to joint wear. This is in contrast to animals with grade IV luxation that do not bear their weight on the affected limb, but drag or carry it while the weight is transferred to the front limbs.

Roy et al. [19] did not find a significant association between the degree of luxation and the progression of radiological articular changes, which suggests that some of the alterations that are visible during surgery might not be detectable in radiological exams.

Because of the anatomy and biomechanics, the friction between the articular surfaces of the patella and femoral trochlea will cause erosion of the medial femoral condyle and lateral patellar surface in medial luxation, and of the lateral femoral condyle and medial patellar surface in lateral luxation [8]. Subluxation also causes wear of the articular surface of the patella and flattening of the condylar edge that favours dislocation [6].

The quadriceps mechanism is responsible extension, and a healthy patellofemoral joint is essential for implementation of this function. The quadriceps, patella, trochlea, patellar tendon and tibial tuberosity must be aligned. Abnormal alignment of the extensor mechanism interferes with limb flexion and extension [8] and compromises its function.

Misalignment of the quadriceps leads the kneecap to become dislocated and press and brush on the lateral or medial condyle surface during limb movement. Erosion is observed in dislocations when the quadriceps' extension function is maintained. In these cases, the pressure of the quadriceps extensor during movement, acting on an improper surface for receiving the force, causes wear injuries. The injury is more severe in animals weighing more than 15 kg. The greater frequency of lesions in grade II and III luxations can be explained by moderate functional changes that allow member extension, promoting compression between the patella and condyle edge. In grade IV luxation, there is a significant decreased range of extension in the joint associated with contracture of the soft tissues caudal to the joint [6, 8] (**Figure 2B, D**), which prevents the friction pressure between the condyle and patellar edge. Patellar luxation is also responsible for the resulting absence of a trochlear groove.

In this study, the frequency and severity of articular lesions were higher among the animals weighing at least 15 kg, as Daems et al. [31] observed. This finding might be attributed to the biomechanical instability resulting from the greater load to which the stifle joint is subjected. Biomechanical stability is considered to be essential for an appropriate supply of blood to the articular cartilage [3], because inadequate nutrition results in joint degeneration, which is observed in cases of patellar luxation.

The high frequency of patellar lesions among animals aged older than 24 months might be the result of disease duration. As some authors have observed [5, 31, 33, 34], patellar luxation mainly affects young animals still in the growing phase, and they are often not referred for treatment for several reasons, among which, the lack of symptoms or the presence of only mild clinical signs, stands out. Consequently, alterations of the patella resulting from chronic friction have already appeared by the time surgery is performed. These are sufficient reasons for early surgical intervention based on the cause—patellar luxation—and not in the presence of clinical signs as reported in the literature [5, 6, 8, 21].

The severity of the long digital extensor tendon injuries, also mentioned in the literature [6], varies according to the chronicity of the case (**Figure 5A, B**) and affects small and large dogs.

As mentioned in the literature [21], severe skeletal changes associated with unilateral patellar aplasia were observed in one cat (**Figure 7A, B**), whose two brothers also had grade III patella luxation and no tail. The litter was a result of crosses between brothers.

The re-luxation rate observed in this study (12.9%) is within the range reported by Arthurs and Langlay-Hobbs [35] and Wandgee et al. [35], although rates up to 50% have been reported [36]. Arthurs and Langley-Hobbs [35] report a greater frequency of major and patellar re-luxation complications in dogs weighing 20 kg or more compared to smaller dogs. In this study, two (6.6%) re-luxations were observed among the 30 dogs that weighed more than 20 kg.

Recurrence of patellar luxation is a common complication associated with surgery; among the factors that contribute to re-dislocation are the severity of the lesions, because grade III and IV luxations indicate poor shaft alignment, failure to align the tuberosity of the tibia with trochlea, and contracture of soft tissues caudal to the joint, affecting the post-operative joint range of motion [6, 8]. As with the loss of convexity of the patella, it loses its slot in the femoral trochlea. Daems et al. [31] postulate that one of the reasons for surgical failure in stable post-surgical patellae is the presence of cartilage erosion. Mostafa et al. [14] proposed that the proximodistal malalignment of the patella (patella alta and baja) influenced re-luxation. Patella alta is defined as the proximal displacement of the patella within the femoral trochlear groove. It has been speculated that patella alta may play a role in canine patellar luxation [12, 37]. The position of the patella in the trochlea, as Mostafa et al. [14] observed, might be the reason for one case of patellar re-luxation in our study. One dog weighing more than 20 kg with recurrent grade II bilateral medial luxation was evaluated. To assess the patellar position on the femoral trochlea, radiographic evaluation was performed according to the method described by Mostafa et al. [13]. After osteotomy of the tibial tuberosity, as described by Johnson et al. [37] and distally transposing 0.5 cm the tibial tubercle, the patella remained in the groove.

We found no instance of implant failure among animals weighing at least 15 kg because the implants used were compatible with the dogs' weights.

The treatment was challenging in both cats with severe deformities of the distal femur, tibia and tail. It was not possible to correct all deformities, but there was improvement in ambulation. Although they were able to move around the house, as related by Hubler et al. [38], the gait abnormalities were extremely severe.

Post-operative measures, such as activity restriction and physical therapy, contribute to the success of treatment and might prevent some post-operative complications and favour an early recovery of muscle mass and the functional performance of the limb.

10. Conclusion

Based on these results, we conclude that patellar luxation should undergo surgical repair even when the clinical signs of disease are mild or absent, because this condition triggers patellar and extra-patellar changes that result in a more difficult surgery and promote the

http://dx.doi.org/10.5772/65764

progression of joint degeneration, thus reducing the odds of a full and painless recovery of joint function.

Acknowledgements

We thank Dr. Oscar Henriques Rocha Ladeira for conducting the photographs.

Author details

Cleuza M.F. Rezende*, Renato César Sachetto Tôrres, Anelise Carvalho Nepomuceno, Juliana Soares Lara and Jessica Alejandra Castro Varón

*Address all correspondence to: cleuzaufmg@gmail.com

Clinical and Surgical Department, Veterinary School of the Federal University of Minas Gerais, Belo Horizonte, Brazil

References

[1] Roush JK. Canine patellar luxation. Vet Clin North Am Small Anim Pract. 1993; 23: 855–868. ISSN: 0023-6772

[2] Nagaoka K, Orima H, Fujita M, Ichiki H. A new surgical method for canine congenital patellar luxation. J Vet Med Sci. 1995; 57: 105–109. doi:10.1292/jvms. 57.105

[3] L'Eplattenier H, Montavon P. Patellar luxation in dogs and cats: pathogenesis and diagnosis. Small Anim Exotics. 2002; 24: 234–240.

[4] Souza MMD, Rahal SC, Padovani CR, et al. Estudo retrospectivo de cães com luxação patelar medial tratados cirurgicamente [Retrospective study of dogs with surgically treated medial patellar luxation]. Cienc Rural. 2010; 40: 31–36. doi:10.1590/S0103-84782010000600016

[5] Piermattei DL, Flo GL, Decamp CE. The stifle joint. In: Piermattei DL, Flo GL, Decamp CE, eds. Handbook of Small Animal Orthopaedics and Fracture Repair. 4th ed. Philadelphia, USA. Saunders, 2006: 562–632. ISBN: 0-7216-5689-7

[6] Denny HR, Butterworth ST. The stifle. In: A Guide to Canine and Feline Orthopaedic Surgery. 4th ed. Iowa. Blackwell Publishing, 2006: 512–553. ISBN: 0-632-05103-5 (hbk)

[7] Evans HE. The skeleton. In: Evans HE, ed. Miller's Anatomy of the Dog. 3rd ed. Philadelphia. WB Saunders Co, 1993: 122–218. ISBN 10: 0721632009 ISBN 13: 9780721632001

[8] McKee MW, Cook JL. The stifle. In: Houlton JEF, Cook JL, Innes JF, Langley-Hobbs SJ, (eds.) BSAVA Manual of Canine and Feline Musculoskeletal Disorders. England, British Small Animal Veterinary Association, 2006: 350-374. ISBN-10 0 905214 80 3, ISBN-13 978 0 905214 80 1

[9] Robins GM. The canine stifle joint. In: Whittick WG, ed. Canine Orthopaedics. 2nd ed. Philadelphia. Lea & Febiger, 1990: 693–760. ISBN: 0812110862

[10] Putnam RW. Patellar luxation in the dog [MS dissertation]. Ontario, Canada: University of Guelph, 1968

[11] Kaiser S, Cornely D, Golder W, et al. The correlation of canine patellar luxation and the anteversion angle as measured using magnetic resonance images. Vet Radiol Ultrasound. 2001; 42: 113–118. doi:10.1111/j.1740-8261.2001.tb00913.x

[12] Johnson AL, Probst CW, De Camp CE, et al. Vertical position of the patella in the stifle joint of the clinically normal large-breed dogs. Am J Vet Res. 2002; 63: 42–46. doi: 10.1111/j.1532-950X.2005.00115.x

[13] Mostafa AA, Griffon DJ, Thomas MW, et al. Proximodistal alignment of the canine patella: radiographic evaluation and association with medial and lateral patellar luxation. Vet Surg. 2008; 31: 201–211. doi:10.1111/j.1532-950X.2008.00367.x

[14] Pérez P, Chelsea WT, Lafuente P. Management of medial patelar luxation in dogs: what you need to know. Vet Ireland J. 2014; 4: 636-640

[15] Kowaleski MP, Boudrieau RJ, Pozzi A. Stifle joint. In: Tobias KM, Johnston SA, eds. Veterinary Surgery Small Animal. St. Louis, Missouri. Elsevier Saunders, 2012: 973–989. ISBN: 978-1-4377-0746-5

[16] Olmstead ML. Lateral luxation of the patella. In: Bojrab MJ, ed. Disease Mechanism in Small Animal Surgery. 2nd ed. Philadelphia. Lea & Febiger, 1993: 818–820. ISBN-10: 0812114914; ISBN-13: 978-

[17] Slocum B, Slocum TD. Alignment problems of the hindlimb. In: Proceedings of 10th ESVOT Congr. 2000: 60–63.

[18] Moller BN, Moller-Larson F, Frich LH. Chondromalacia induced by subluxation in the rabbit. Acta Orthop Scand. 1989; 60: 188–191. doi:10.3109/17453678909149251

[19] Roy RG, Wallace LJ, Johnston GR, et al. A retrospective evaluation of stifle osteoarthritis in dogs with bilateral medial patellar luxation and unilateral surgical repair. Vet Surg. 1992; 21: 475–479. doi:10.1111/j.1532-950X.1992

[20] Singleton WB. The surgical correction of stifle deformities in the dog. J Small Anim Pract. 1969; 10: 59–69. doi:10.1111/j.1748-5827.1969.tb04021.x

[21] Houlton JEF, Meynink SE. Medial patellar luxation in the cat. J Small Anim Pract. 1989; 30: 349–352. doi:10.1111/j.1748-5827.1989...x

[22] Engvall PD, Bushnell N. Patellar luxation in Abyssinian cats. Feline Pract. 1990; 18: 20–22.

[23] Flecknell PA, Gruffydd-Jones JJ. Congenital luxation of the patella in the cat. Feline Pract. 1979; 9: 18–20.

[24] Smith GK, Langenbach A, Green PA, Rhodes WH, Gregor TP, Giger U. Evaluation of the association between medial patellar luxation and hip dysplasia in cats. J Am Vet Med Assoc. 1999; 215: 40–45.

[25] Johnson ME. Feline patellar luxation: a retrospective case study. J Am Anim Hosp Assoc. 1986; 22: 835–838. https://www.aaha.org/professional/resources/jaaha.aspx

[26] Langley-Hobbs SJ. Patellar luxation–what is different between cats and dogs. In: Proceedings of 22. FECAVA Eurocongress 31st Annual Congress of the Association of Austrian small animal veterinarians. Orthopedic Surgery Hofburg, Vienna, June 2016: 22–25.

[27] Voss K, Langley-Hobbs SJ, Montavon PM. Stifle joint. In: Montavon PM, Voss K, Langley-Hobbs SJ, eds. Feline Orthopedic Surgery and Musculoskeletal Disease. Edinburgh. Elsevier, 2009: 475–490. www.ncbi.nlm.nih.gov/pubmed/16649941

[28] Loughlin CA, et al. Clinical signs and results of treatment in cats with patellar luxation: 42 cases. J Am Vet Med Assoc. 2006; 228: 1370–1375. www.ncbi.nlm.nih.gov/pubmed/16649941

[29] Milovancev M, Rhalphs SC. Congenital patellar aplasia in a family of cats. Vet Orthop Traumatol. 2004; 17: 9–11. doi:10.3415/VCOT-07-10-0092.

[30] Remedios AM, Basher AWP, Runyon CL, et al. Medial patellar luxation in 16 large dogs: a retrospective study. Vet Surg. 1992; 21: 5–9. doi:10.1111/j.1532-950X.1992

[31] Daems R, Janssens LA, Béosier YM. Grossly apparent cartilage erosion of the patellar articular surface in dogs with congenital medial patellar luxation. Vet Comp Orthop Traumatol. 2009; 22: 222–224. doi:10.3415/VCTO-07-08-0076

[32] Vasseur PB. Stifle joint. In: Slatter D, ed. Textbook of Small Animal Surgery. 3rd ed. Philadelphia, USA. Saunders, 2003: 2090–2133. ISBN 0-7216-8607-9

[33] Hayes AG, Boudrieau RJ, Hungerford LL. Frequency and distribution of medial and lateral patellar luxation in dogs: 124 cases (1982–1992). J Am Vet Med Assoc. 1994; 57: 105–109. www.ncbi.nlm.nih.gov/pubmed/7989241

[34] Gibbons SE, Macias C, Tonzing MA, Pinchbeck GL, Mackee WM. Patellar luxations in 70 large breed dogs. J Small Anim Prac. 2006; 47: 3–9. doi:10.1111/j.1748-5827.2006.00004.x

[35] Arthurs GI, Langley-Hobbs SJ. Complications associated with corrective surgery for patellar luxation in 109 dogs. Vet Surg. 2006; 35: 559–566. doi:10.1111/j.1532-950X.2006.00189.x

[36] Wandgee C. Evaluation of surgical treatment of medial patellar luxation in pomeranian dogs. Vet Comp Orthop Traum. 2013; 26: 435–439. doi:10.3415/VCOT-12-11-0138

[37] Johnson AL, Broaddus KD, Hauptman JG, et al. Vertical patellar position in large-breed dogs with clinically normal stifle and large-breed dogs with medial patellar luxation. Vet Surg. 2006; 35: 78–81. 10.1111/j.1532-950X.2005.00115.x

[38] Hubler M, Arnold S, Langley-Hobbs. Hereditary and congenital musculoskeletal diseases. In: Montavon PM, Voss K, Langley-Hobbs SJ, eds. Feline Orthopedic Surgery and Musculoskeletal Disease. Edinburgh. Elsevier, 2009: 41–53. ISBN: 978-0-7020-2986-8

Section 6

Diagnostic Neurology

Chapter 9

Diffusion Tensor Tractography in Cerebral White Matter

Mitzi Sarahi Anaya García and
Jael Sarahi Hernández Anaya

Additional information is available at the end of the chapter

http://dx.doi.org/10.5772/66249

Abstract

Conventional magnetic resonance imaging (MRI) allows researchers and clinicians to observe the anatomy and injuries of the cerebral white matter (CWM) in dogs. However, dynamic images based on the diffusion tensor (DT) technique are required to assess fiber tract integrity of the CWM. Diffusion tensor tractography (DTT) produces a three-dimensional representation in which data are displayed on a colored map obtained from the anisotropy of water molecules in the CWM tracts. Fractional anisotropy (FA) is a value that measures changes in water diffusion, which can occur if the CWM tracts are displaced, disrupted, or infiltrated. The goal of this study was to determine the feasibility of DTT for in vivo examination of the normal appearance of CWM in dogs through visual and quantitative analysis of the most representative CWM tracts.

Keywords: diffusion tensor, cerebral white matter, dog, tractography, in vivo

1. Introduction

Magnetic resonance imaging (MRI) is the best imaging technique for the evaluation of the neurologic system due to the ability to diagnose central nervous system (CNS) diseases in both humans and animals, especially diseases affecting cerebral white matter (CWM), providing in vivo markers for disease severity or response to therapy and shedding light on progression and recovery processes. The broad spectrum of magnetic resonance contrast mechanisms makes MRI one of the most important and widely used imaging tool for diagnosis in the CNS. However, one of the disadvantages of conventional MRI is that it does not allow the visualization of the cerebral White matter constituent fiber tracts and their connectivity in the brain in vivo [1].

Previously, the anatomy of the cerebral white matter tracts could only be studied by postmortem dissection or through invasive methods, and said methods could only reveal a few tracts in vivo (neurosurgery) and no tracts could be observed in vivo via conventional imaging studies [1–6].

Diffusion tensor tractography (DTT) is a useful noninvasive imaging technique that can identify and represent fiber tracts of the cerebral white matter and their connections in the brain in vivo [1]; also, it can give us information that cannot be achieved by conventional anatomical MRI or histology. In addition to displaying specific cerebral white matter fiber tracts, this technique can also improve the quantification of diffusion characteristics within these fibers [1, 7–9].

DTT provides a three-dimensional representation of diffusion tensor imaging (DTI), and data can be displayed on a colored map obtained from information on the directionality of the movement of water molecules along the main fiber tracts of cerebral white matter [1].

1.1. Diffusion tensor imaging

DTI uses the property of the water diffusion anisotropy in axonal fibers allowing the analysis and tracking of said fibers in the brain [1, 4, 10, 11].

DTI permits the exploration of microstructural tissue features through the observation of water molecular diffusion, thus furnishing information about the anatomy, microstructural features, and damage of the main brain bundles, useful in several pathological animal models. DTI-based tractography permits the virtual reconstruction of the white matter fiber bundles in-vivo, following the principal diffusion direction [12, 13].

This technique is commonly used in human medicine to study the anatomy and maturation of the normal, aging brain, but it also can be used to help diagnose neurological conditions, including brain ischemia, multiple sclerosis, diffuse axonal injury, epilepsy, metabolic disorders, certain mental illnesses, and brain tumors, as well as establish a prognosis for patients with these conditions [14, 15]. Though DTI has been extensively used to investigate brains of the dogs ex vivo [11]; there is only one report of the in vivo use of DTT to study cerebral white matter fiber tracts in dogs [1].

The ability to trace cerebral white matter fibers in the dog generates a number of opportunities for potential clinical applications, and has both diagnostic and prognostic [1].

2. Diffusion anisotropy measurement and tensor analysis

Diffusion is a physical process that involves the translational movement of molecules via thermally driven random motions, the so-called Brownian motion. The factors influencing diffusion in a solution (or self-diffusion in a pure liquid) are molecular weight, intermolecular interactions (viscosity), and temperature [16, 17]. Diffusion is a random transport phenomenon, which describes the transfer of material from one spatial location to other locations over time. The direction of water molecules diffusion in living tissues is always limited to some degree. Water diffusion in biological tissues occurs inside and outside of cellular structures, and it is caused primarily by random thermal fluctuations. The water diffusion is affected by the interaction with cellular and subcellular membranes and with organelles. Cellular membranes deter water diffusion, decreasing the water mean squared displacement. The diffusion hindering and corresponding apparent diffusivity may increase by either cellular swelling or increased cellular density [3].

On the other hand, breakdown of cellular membranes caused by necrosis or other ailments increases the apparent diffusivity. Intracellular water tends to be more contained by cellular membranes, rather than deterred. This restricted diffusion also decreases the apparent diffusivity, but plateaus with increasing diffusion time [3]. In fibrous tissues, including white matter, water diffusion is relatively unimpeded in the parallel direction to the fiber orientation. On the contrary, water diffusion is highly restricted and deterred in directions perpendicular to the fibers. Hence, the diffusion in fibrous tissues is anisotropic [3, 18].

The property by which the rate of diffusion varies with direction is called diffusion anisotropy or anisotropic diffusion [4]. Isotropic diffusion occurs when the magnitude of diffusion is the same in all directions. Conversely, anisotropic diffusion is when the magnitudes of diffusion are significantly different [19].

In some tissues, for example cerebrospinal fluid (CSF), Brownian motion leads water molecules to diffuse freely in any direction. For other tissues, like white matter, water diffusion occurs along the fiber orientation rather than across it due to the highly organized fibrous structure that restricts water diffusion [20]. The underlying tissue cellular microstructure influences the overall mobility of the diffusing molecules by providing numerous barriers and by creating various individual compartments (e.g., intracellular, extracellular, neurons, glial cells, axons) within the tissue [16]. Early diffusion imaging experiments used measurements of parallel and perpendicular diffusion components to characterize the diffusion anisotropy [3].

The behavior of the anisotropic diffusion using diffusion tensor (DT) is described by a multivariate normal distribution, which describes the covariance of diffusion displacements in three dimensions normalized by the diffusion time [3]. Water diffusion cannot be characterized by a single value in an anisotropic voxel given its directional dependence; thus, the tensor model was developed. A tensor may assist in obtaining different parameters. For example, a three-dimensional principal eigenvector indicates the gradient of water diffusion within a voxel. In a similar manner, scalar eigenvalues signify the magnitude of the diffusivities along the principal and two orthogonal eigenvectors [21]. The diagonal elements are the diffusion variances along the axes x, y, and z, and the off-diagonal elements are the covariance terms and are symmetric about the diagonal [21]. The diffusion tensor can be represented as an ellipsoid, with its principal axis being defined by the eigenvectors and the ellipsoidal radii defined by the eigenvalues. If the eigenvalues are nearly equal, the diffusion is considered isotropic; if significantly different, anisotropic. Local tissue microstructure modifications such as injury, disease, or normal physiological changes may alter the eigenvalue magnitudes. Hence, the diffusion tensor allows the characterization of both normal and abnormal tissue microstructure [3]. In the CNS, water diffusion is usually more anisotropic in white matter regions and isotropic in both grey matter and cerebrospinal fluid (CSF). The major diffusion eigenvector is assumed to be parallel to the tract orientation in regions of homogeneous white matter [3].

Diffusion-weighted imaging (DWI) is an important technique of functional magnetic resonance (fMR) imaging, which has the ability to assess changes in random motion of water protons in vivo. It is useful to diagnose several diseases in the central nervous system of humans and animals, specially canines [2, 22]. To detect lesions by DWI, the anisotropy is deliberately reduced using imaging techniques and processing to avoid detecting the signals from normal white matter [4].

DWI can be used to detect and visualize water molecules diffusion in tissues by adding a bipolar gradient pulse called a motion-proving gradient (MPG). Diffusion-weighted MRI differs from conventional MRI in that it provides high-contrast resolution based on diffusion, which allows new information on lesions to be obtained [4]. In the 1970s, water diffusion MRI was introduced and later used for medical applications [5]. Reports on diffusion MRI of the brain for neurological disorders were first published in the 1980s [23]. In the 1990s, its use was extended [4]. With the introduction of DTI, it was proposed to represent the water diffusion coefficient distribution in all the directions of space as a tensor in each voxel. A reconstruction of the white matter pathways was later proposed based on this tensor model [5].

DTI is an advanced technique of DWI sequence that displays vectors corresponding to the strength and direction of the movement of water molecules [2]. Recently, the DTI technique has permitted the detailed visualization of white matter structural integrity and connectivity [24]. One of the advantages of DTI is the reconstruction of axonal tracts in the brain in vivo [10]. DTI uses water diffusion anisotropy in axonal fibers allowing the analysis and tracking of said fibers in cerebral white matter.

Cerebral white matter anatomy can be studied in detail using DTI; it shows a complete anatomical and statistical fiber atlas of the white matter [15]; and it can explain, in combination with functional MRI, some anatomical and functional connectivity between different parts of the brain [11].

Pathologic conditions such as edema, inflammation, myelin loss, and gliosis may cause disruption in white matter tracts or changes in the membrane permeability which can alter DTI measurements, such as fractional anisotropy (FA) and apparent diffusion coefficient (ADC) [25].

Several studies have demonstrated the validity of quantitative diffusion imaging of the large white matter tracts in the brain in vivo [2]. FA provides information about the shape of the diffusion tensor at each voxel. The FA relates the differences between isotropic and anisotropic diffusion and is a scalar value between 0 and 1 indicating the degree of anisotropy in water diffusion. If FA value is close to 0, the diffusion is isotropic or random, and if it is close to 1, the diffusion is highly directional. The diffusion coefficient measured by nuclear magnetic resonance is best known as apparent diffusion coefficient (ADC). ADC depends greatly on the interactions of the diffusing molecule with the cellular structures over a given time; it could also be influenced by active processes within the tissue. ADC is calculated by acquiring two or more images with a different gradient duration and amplitudes, quantified as b-values.

http://dx.doi.org/10.5772/66249

Also, one can calculate the eigenvalues corresponding the various imaging axis in order to find if the water diffusion is either anisotropic or not. With the obtained data, a tensor map is generated and tractography can be performed. This entire process is called DTI [2, 3, 16, 25, 26]. DWI uses an ADC map; given that DTI is an extension of DWI, it also uses an ADC map [16]. The difference between DWI and DTI is that in DWI encodes water diffusion in three spatial directions and DTI uses up to six directions.

The strength of the signal in FA maps represents the magnitude of FA, and, unlike conventional DWI, it is quantitative. FA decreases at lesion sites. Diffusion tensor analysis has the novel feature of not only quantifying anisotropy by FA and other parameters, but also of analyzing directionality.

Colored FA maps represent anisotropy in different colors according to the direction of the principal axis. These maps make it possible to differentiate fibers based on the direction in which they run. The colors are assigned to nerve fiber tracts depending on the direction of water displacement. The colors represent the predominant orientation of the fibers in a three-dimensional coordinate system in the three axes of space (x, y, and z); where red indicates a right-left direction, green indicates a dorsoventral direction, and blue indicates a rostrocaudal direction. This type of representation is called an anisotropic [1, 4, 27].

2.1. Construction of diffusion tensor tractography (DTT)

Methods for tracing connections in the brain have a long history, beginning with those based on lesions and the resulting retrograde or anterograde degeneration. The ensuing methods exploited the axonal transport of specific molecules like the horseradish peroxidase, and it was followed by a host of other tracers including small fluorescent molecules, lectins, neurotrophins, neurotoxins, dextrans, carbocyanine dyes, latex microspheres, and viruses. Although these methods allow the study of the brain connections, they are highly invasive. Moreover, any histological visualization of the transported substance requires sacrifice of the experimental animal [5]. Magnetic resonance images that make use of tensor analysis, such as FA maps and color maps, are collectively called DTI. A recent extension of DTI is fiber tracking, or tractography, which has been applied in the brain to noninvasively identify specific white matter pathways and connections in the brain in vivo. In the broadest sense, DTT can be considered a subtype of DTI, but it is often deliberately differentiated from DTI.

DTI and DTT provide us with a new opportunity to investigate such structures and to assess changes due to brain disease [4, 7, 15, 27]. DTT is a method of noninvasively tracing neuronal fiber bundles, and it integrates voxel-by-voxel orientations into a pathway that connects distant brain regions. The DTT technique can be used to analyze the trajectory, shape, fiber structure, location, topology, and connectivity of neuronal fiber pathways in vivo [28–30]. DTT can be used to simultaneously delineate cerebral white matter tracts in three dimensions and to identify alterations in connectivity. By tracing fiber pathways throughout the entire brain, diffusion tractography provides information that cannot be achieved by conventional anatomical MR imaging or histology [8, 9]. DTT refers to 3D models of white matter pathways generated from diffusion weighted MRI data, most commonly diffusion tensor imaging (DTI). Here, the term DT is used to refer to all forms of tractography derived from diffusion MRI data including but not limited to DTI. Given that white matter is highly anisotropic, the nerve fibers can be tracked and visualized with DTT. To perform a DTT, parameters such the ADC and the FA must be calculated. To calculate these parameters, tensor analysis is employed [4, 27, 31]. The orientation of the component of the diagonalized diffusion tensor represents the orientation of the dominant axonal tracts, DTI provides a 3D vector field, and each vector represents the fiber orientation. Actually, there are different ways to reconstruct cerebral white matter tracts, these reconstructions are divided in two different types. The first type is based on line propagation algorithms that use local tensor information for propagation step. The main differences found throughout these techniques are due to the way they incorporate information from neighboring pixels in order to define smooth trajectories and to reduce noise contributions. The second one is based on global energy minimization to find the most favorable energetic pathway between two pixels [15]. The orientation of the diffusion tensor major eigenvector is generally assumed to be parallel to the local white matter fascicles. These directional patterns may be simply visualized using the color maps representing the major eigenvector direction. Such color maps are useful for assessing the organization of the cerebral white matter in the brain and for the identification of the major cerebral white matter tracts in 2D sections [3]. Another approach to visualize the white matter connection patrons in 3D is using diffusion tensor tractography. White matter patterns are estimated by starting at a specified location, this location is called seed point, also estimating the direction of propagation that is called major eigenvector, and moving a short distance in that direction called tract integration. The tract direction is then re-evaluated and another small step is taken until the tract is finished. Fiber tracts can be obtained using different number of regions of interest (ROIs) [3]. Diffusion tensor tractography have been used to produce anatomically fiber tract reconstruction of the most important proyection pathways. The primary applications of tractography to date have been the visualization of WM trajectories in 3D (particularly, in relation to brain pathology) and segmentation of specific brain regions [3, 32]. In human medicine, this technique has great potential for studying numerous diseases because the results of DTT and clinical symptoms can be compared directly. This technique is commonly used in human to study the anatomy and maturation of the normal, aging brain, but it also can be used to help diagnose neurological conditions, including brain ischemia, multiple sclerosis, diffuse axonal injury, epilepsy, metabolic disorders, certain mental illnesses, and brain tumors, as well as establish a prognosis for patients with these conditions [1]. FA and ADC are measurements that may be altered by disruption of white matter tracts or changes in membrane permeability as a result of pathologic conditions such as edema, inflammation, myelin loss, and gliosis. We can visualize the physical displacement of axon bundles employing tractography maps and are evaluated for surgical planning. Also, using axonal diffusivity, we can quantified noninvasively and correlate with histopathology axonal injury without demyelination [1, 2, 11, 25, 33].

2.2. DTI in human medicine

DTI is a widely used technique for the detection of several central nervous system diseases. It is the tool of choice because it is highly sensitive, highly specific, and noninvasive while providing a diagnosis within the therapeutic window [34].

It is expected to become even more useful due to the recent development of diffusion-weighted whole-body imaging with background body signal suppression, which can be used to evaluate the metastases of malignant tumors in three dimensions.

DTI is commonly used in order to study the normal anatomy of normal brain maturation and aging [11].

Normal white matter brain anatomy can be studied in detail using DTI; on one side, it shows a complete anatomical and statistical fiber atlas of the white matter, and on the other hand, it can explain in combination with functional MRI, some anatomical and functional connectivity between different parts of the brain [11, 15].

Aside detecting diseases in the central nervous system, DTI has been used for different applications in human medicine.

DTI was applied presurgically in human medicine to plan function-preserving brain surgery by saving special areas of motor and speech function, actually the application of the DTI is also possible for white matter tract reconstruction in 3D images for the spinal cord and brain [22].

Some applications of DTI in normal brain are demonstrating the relationship between the white matter structure and its function; for example, IQ has been positively correlated with anisotropy in cerebral white matter association tracts. Reading ability has been correlated with anisotropy of the left temporoparietal white matter, where tractography has localized the language areas. In visual pathways, increase in anisotropy has been correlated with improved reaction time. Tractography findings have demonstrated an excellent correlation with functional data; for example, probabilistic tractography has been used for segmentation of the thalamus according to its cortical connectivity, which corresponds well to segmentation of the thalamus.

DTI has shown additional abnormalities in patients with several types of dementia and neurodegenerative disease.

Several researches have use DTI to demonstrate a variety of white matter abnormalities often correlated with performance in neuropsychiatric test.

For demyelinating disease, the use of DTI is often used for diagnosing multiple sclerosis; several groups have demonstrated increased diffusivity and decreased anisotropy in demyelinating lesions.

In the case of ischemic disease, DTI is used for the detection of early acute ischemia into the domain of prognosis and long-term management of ischemic sequelae.

DTI has important implications in the delineation of tumor margins beyond what is currently demonstrated with conventional MRI [21].

2.3. DTI in animals

DTI has been used in an experimental model in cats, rats, mice, dogs, pigs, and marmosets [2, 5, 8–10, 12, 17, 20, 30, 31, 34]. DTI has generally been used following the creation of spinal cord trauma to serve as a model to aid in evaluating human spinal cord trauma victims [4, 35–37].

In dogs, it has been used for the evaluation of the spinal cord and its pathologies, providing statistically and visually different images when evaluating fractional anisotropy and ADC in normal dogs compared against dogs with lesions localized in the spinal cord using different types of scanners and software [19, 22, 24, 38–40].

DTI technique has also been applied to image animal brains in vivo and ex vivo [1, 10, 11, 20, 34]. Previous work shows consistent results between diffusion anisotropy of in vivo and ex vivo formalin-fixed mouse brains. This offers a new opportunity to study the brain micro-structure with ex vivo DTI, as it avoids motion artifacts and allows for longer imaging time [20].

The use of DTI in white matter tracts found in dog and human brains has the potential for studying several pathologies by correlating DTI findings with clinical symptoms [11].

The corpus callosum has been also studied in dogs with DTI because its aging and pathologies are similar of those found in the human corpus callosum. In Pierce's work, the corpus callosum was segmentated into six major White matter tracts, which will provide grounds for new research in both species [34].

Tracing tracts by DTI is relative new, and it is expected that further research will develop new technologies soon. However, recent studies have demonstrated that even simple methodologies are able to visualize cerebral white matter tracts connections in situ for both human and animals [15].

2.4. DTI limitations

To date, DT remains the only noninvasive method for visualizing human brain and spinal cord connections. DT suffers from both fundamental and practical limitations that limit its use for modelling brain connections. Unlike many invasive modalities, DT is incapable of determining the direction of information flow, nor can it distinguish single- and multineuron connections. DT may also have difficulty in resolving complex intravoxel fiber crossings or nondominant fiber populations due to limitations in scan time, hardware, or processing methods. Despite its many limitations, DT has been successfully used to model human neuronal connections for over two decades, including several pathways that are putatively deep brain stimulation targets. DT generation can be divided into three separate steps: data acquisition, data processing, and tracking. Each of these steps has several variables that must be considered in order to ensure accurate DT [41].

2.5. New research on DTI

Different methods for the acquisition and analysis of DTI have been developed and have improved the precision of diffusion tensor measurements in recent years, so, new innovations can be expected. New pulse sequences and diffusion tensor encoding schemes are being developed to improve the spatial resolution, accuracy and to decrease artifacts in diffusion tensor measurements [3].

Even though DTI provides quantitative parameters of clinical relevance, it is limited in representing complex diffusion schemes. Methods based on high angular resolution diffusion imaging (HARDI) provide a more precise diffusion profile visualization [36].

Q-ball imaging (QBI) allows the detection of subtle anatomical features of the spinal cord that were not seen with DTI.

QBI has also been applied to the injured spinal cord, demonstrating its ability to detect directional abnormalities. Metrics derived from QBI may therefore provide useful markers of diffusion characteristics in the healthy and injured SC [36, 37].

3. Diffusion tensor tractography of cerebral white matter in the dog

Several studies have demonstrated the validity of quantitative diffusion imaging of the large white matter tracts in the brain in vivo [25]. DTI has been extensively used to investigate brains of the dogs ex vivo [11, 34].

There is only one report of the use of DTT in dogs to determinate the feasibility for in vivo examination of the normal appearance of the cerebral white matter.

MRI allows investigators and clinicians to observe the anatomy and injuries of the cerebral white matter (CWM) in dogs. However, dynamic images based on the diffusion tensor (DT) technique are required to assess fiber tract integrity of the CWM. The goal of this study was to determine the feasibility of DTT for in vivo examination of the normal appearance of CWM in dogs through visual and quantitative analysis of the most representative CWM tracts [1].

3.1. Materials and methods: experimental animals

Nine healthy canine patients of varying breeds and genders were prospectively recruited for the study. The dogs received a general physical and neurological exam, and blood samples were taken for a preanesthetic profile. During the imaging procedure, the dogs were anesthetized with diazepam (2 mg/kg IV, valium, Roche, Nutley, New Jersey) and propofol (4 mg/kg IV, recofol, Bayer, Turku, Finland) [1].

3.2. Image acquisition technique

The MRI protocol was carried out in the same 3 Tesla scanner for all dogs. Diffusion tensor imaging was performed on each patient. Moreover, T1- and T2-enhanced images were acquired to obtain a high-resolution anatomical reference. T1- and T2-weighted images and DTI were obtained in different planes (transverse, dorsal, and sagittal). Three-dimensional reconstructions, FA, and ADC values were obtained for the left and right corticospinal tracts, the corpus callosum, the cingulum, and the right and left fronto-occipital fasciculus to visually evaluate and quantify these fiber tracts [1].

3.3. Diffusion tensor imaging tractographies

Diffusion tensor tractography was performed by importing DTI into image analysis software. Cerebral white matter tracts were identified using regions of interest (ROIs). The software identified tracts based on finding the most favorable path between two manually placed ROIs. Regions of interest were positioned where trajectories of the cerebral white matter fiber tracts were estimated to be, based on veterinary anatomy guides and a human DTI atlas. High-resolution T1-images were placed on top of the colored map to identify connections between anatomical structures. The different tracts were identified, delineated, and reconstructed at different points along their trajectory using the color map in the sagittal, dorsal, and transverse planes, which were reconstructed using a fiber-tracking algorithm. Data were coded in red to indicate a right-left direction, green to indicate a dorsoventral direction, and blue to indicate a rostrocaudal direction. The cerebral white matter tracts were assigned to three groups of fibers: projection, commissural, and association fibers [1].

3.4. Data analysis

Statistical analyses were selected and performed using a commercially available statistical software package (SSPS, version 19, Microsoft, Chicago, IL). Mean tract FA and ADC values, their standard errors, and standard deviation were calculated. A confidence interval of 95% or a significance value of $P < 0.05$ was used for the mean. A quantitative assessment of ventricular volume (VV) in relation to the brain volume (BV) was also performed using manual segmentation in regions of interest (ROIs) on the image analysis freeware (OsiriX v.3.9.4) in the nine healthy dogs. The means, standard errors, standard deviations, and 95% confidence intervals for the means of the VV in relation to the BV of the right and the left side were obtained [1].

3.5. Results

Three-dimensional reconstructions of the corticospinal tract, corpus callosum, cingulum, and fronto-occipital fasciculus were generated for each of the nine dogs. Fibers in the corticospinal tract component of the projection fiber group were displayed in blue and green (**Figure 1A–C**) [1].

Blue fibers connected cortical areas in the cerebral cortex, the brain stem, and spinal cord. Green fibers connected the corona radiate, internal capsule, and cerebral peduncle. Fibers in the corpus callosum component of the commissural fiber group were displayed in red and connected the two cerebral hemispheres (**Figure 2A–C**).

The cingulum component of the association fiber group appeared as long fibers, and these were displayed in blue (**Figure 3A–C**) [1].

These fibers had a rostrocaudal orientation and connected cortical areas in each hemisphere. Fibers in the superior and inferior fronto-occipital fasciculus component of the association fiber group were long and displayed in blue (**Figure 4A–C**) [1].

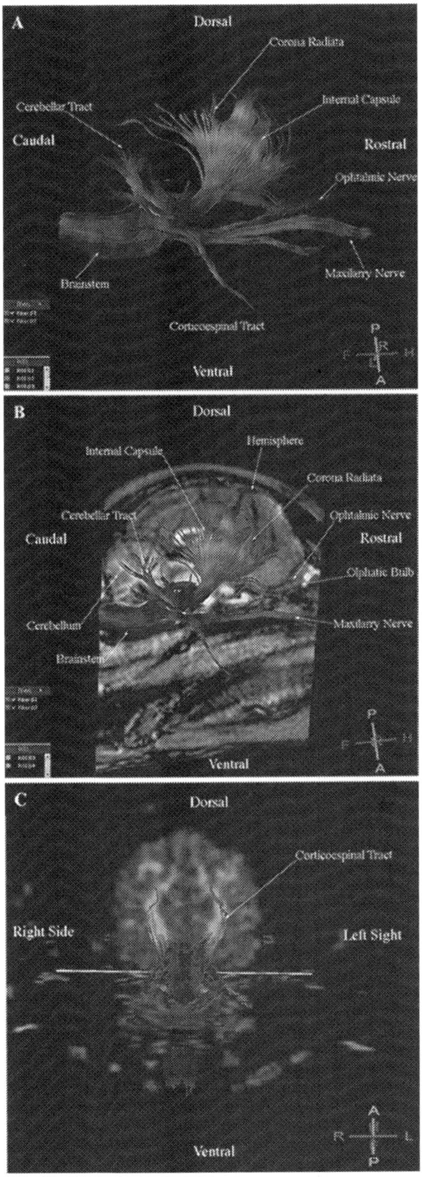

Figure 1. Images illustrating the corticospinal tract in a healthy dog. (A) DTT image, side view. (B) T1-weighted image, sagittal view. (C) Colored map, transverse sectional view [1].

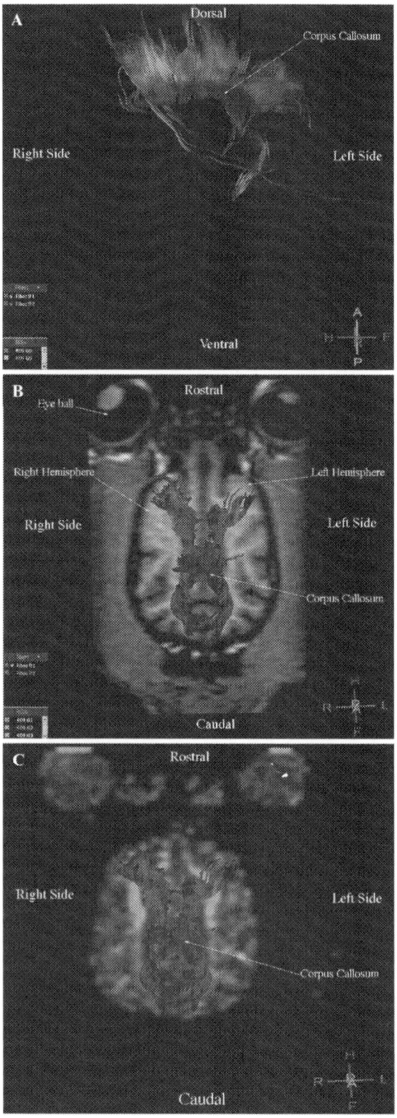

Figure 2. Images illustrating the corpus callosum in a healthy dog. (A) Diffusion tensor tractography (DTT) image, side view. (B) T1-weighted image and diffusion tensor tractography (DTT) image, dorsal view. (C) Diffusion tensor tractography (DTT) image on the colored map, dorsal view [1].

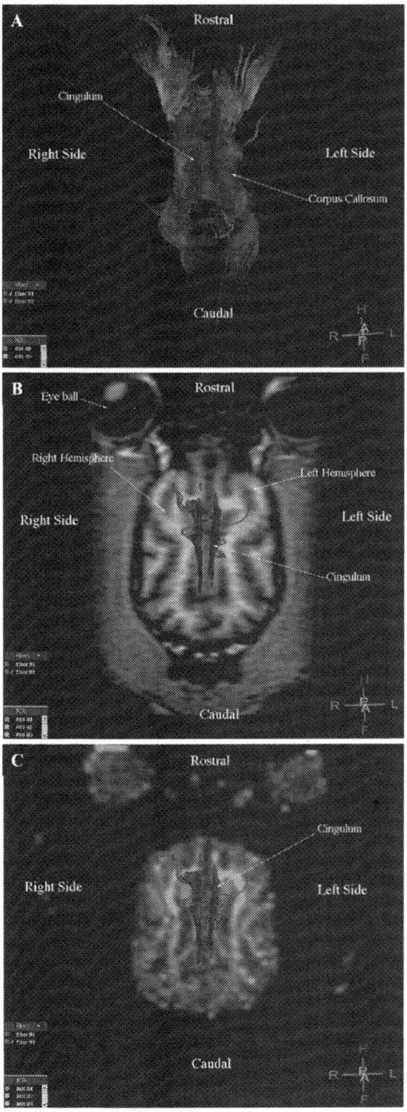

Figure 3. Images illustrating the cingulum in a healthy dog. (A) Diffusion tensor tractography (DTT) image, dorsal view. (B) T1-weighted image and diffusion tensor tractography (DTT) image, dorsal view. (C) Diffusion tensor tractography (DTT) image on the colored map, dorsal view [1].

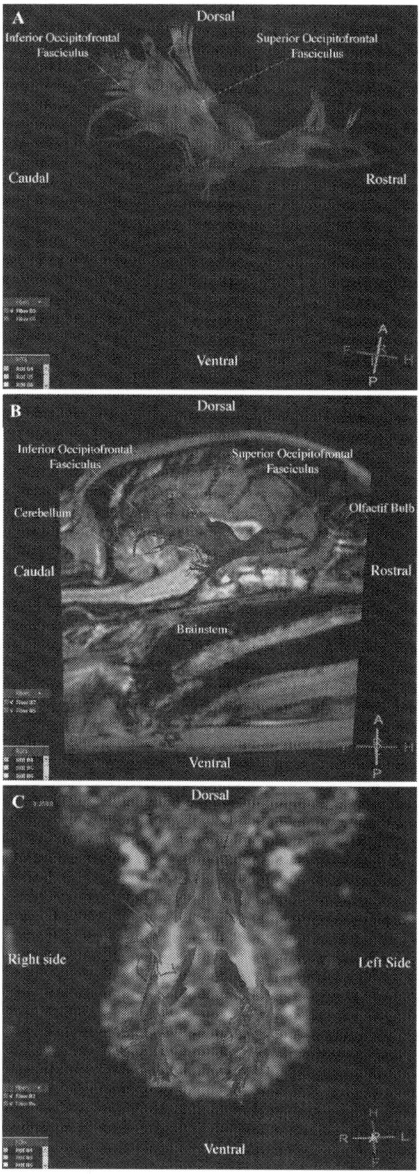

Figure 4. Images illustrating the fronto-temporo-occipital tract in a healthy dog. (A) Diffusion tensor tractography (DTT), side view. (B) T1-weighted image and diffusion tensor tractography (DTT) image, sagittal view. (C) Diffusion tensor tractography (DTT) image on the colored map, dorsal view [1].

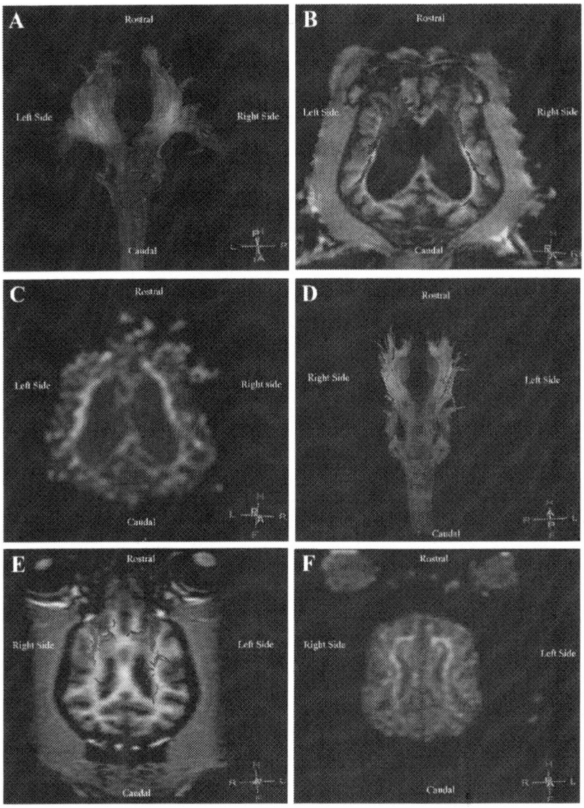

Figure 5. Comparison between a healthy dog with large lateral ventricles and a healthy dog with normal ventricles. (A) Diffusion tensor tractography (DTT) image, corticospinal tract with altered topography and corpus callosum due to large lateral ventricles, dorsal view. (B) Diffusion tensor tractography (DTT) image, corticospinal tract with normal downward topography of the fibers, dorsal view. (C) T1-weighted image and diffusion tensor tractography (DTT) image, corticospinal tract with anterior and lateral displacement of the fibers and of the corpus callosum due to large lateral ventricles, dorsal view. (D) T1-weighted image and diffusion tensor tractography (DTT) image, corticospinal tract with normal downward topography of the fibers, dorsal view. (E) Diffusion tensor tractography (DTT) image of the corticospinal tract on the colored map with anterior and lateral displacement of the fibers and of the corpus callosum due to large lateral ventricles, dorsal view. (F) Diffusion tensor tractography (DTT) image, corticospinal tract on the colored map with normal topography of the fibers, dorsal view [1].

These fibers had a rostrocaudal orientation and connected cortical areas in each hemisphere and in the frontal and occipital lobes. Three-dimensional reconstructions of the tracts were homogeneous and uniform in geometry and spatial orientation in eight of the nine healthy dogs. In one dog, tract reconstructions were not homogeneous or uniform because the fibers were displaced, most evident in the corticospinal tract and corpus callosum (**Figure 5**) [1].

There was a significant difference in the VV in relation to the BV in this dog (**Table 1**) [1].

Relation ventricle volume/brain volume	Healthy dogs (n = 9)			
	X	SE	SD	C.I. 95%
RVV/BV (left)	0.051	0.019	±0.057	(0.007, 0.095)
RVV/BV (right)	0.044	0.016	±0.047	(0.007, 0.080)

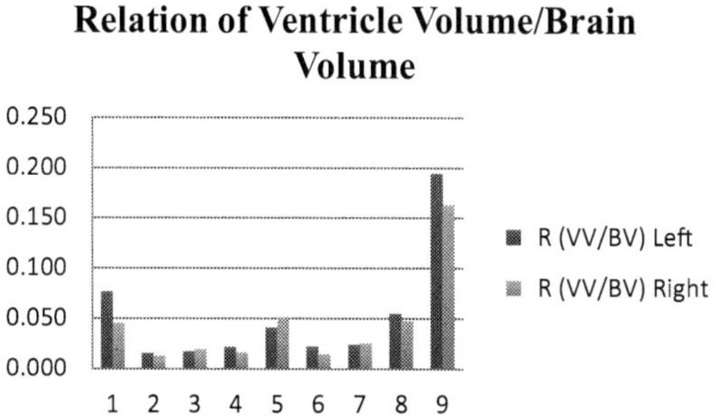

R, relation; VV, ventricle volume; BV, brain volume; X, mean; SE, standard error; SD, standard deviation; C.I., 95% confidence interval for the mean.

Table 1. Ventricular volume (VV) in relation to the brain volume (BV) in healthy dogs <inlinefx>.

The means, standard errors, standard deviations, and 95% confidence intervals for the means of the FA and ADC values for the six tracts were obtained from all nine dogs (**Table 2**) [1].

Tracts	Healthy dogs (n = 9)							
	FA				ADC (10^{-3} mm^2 s^{-1})			
	X	SE	SD	C.I. 95%	X	SE	SD	C.I. 95%
Corticospinal (right)	0.391	0.009	±0.026	(0.371, 0.412)	1.154	0.063	±0.189	(1.009, 1.299)
Corticospinal (left)	0.400	0.006	±0.018	(0.387, 0.414)	1.104	0.055	±0.164	(0.977, 1.230)
Corpus callosum	0.365	0.007	±0.022	(0.348, 0.382)	0.975	0.033	±0.099	(0.899, 1.051)
Cingulum	0.336	0.011	±0.032	(0.311, 0.360)	0.847	0.053	±0.159	(0.725, 0.969)
Fronto-occipital (right)	0.331	0.009	±0.026	(0.311, 0.350)	0.879	0.009	±0.028	(0.858, 0.901)
Fronto-occipital (left)	0.331	0.010	±0.029	(0.308, 0.353)	0.913	0.010	±0.031	(0.888, 0.937)

FA, fractional anisotropy; ADC, apparent diffusion coefficient; X, mean; SE, standard error; SD, standard deviation; C.I., 95% confidence interval for the mean.

Table 2. Fractional anisotropy and apparent diffusion coefficient values of six cerebral white matter tracts in healthy dogs.

Similarities in the FA and ADC values were identified in the nine healthy dogs [1].

http://dx.doi.org/10.5772/66249

3.6. Discussion

The current study was the first to visually and quantitatively describe the trajectory of cerebral white matter fiber tracts in a group of live dogs using DTT for diagnostic purpose. Diffusion tensor tractography-imaging resolution allowed rapid display and identification of the most representative cerebral white matter fiber tracts in this sample population of nine healthy dogs of varying breeds and genders. This technique described anatomical, geometric, and spatial properties of the fibers. In addition, the conduction properties of the fibers could be estimated through FA and ADC quantification. There was homogeneity and uniformity in the three-dimensional reconstructions in nearly all dogs. Quantification of the FA and ADC values of the most representative tracts was similar in all nine dogs, thus demonstrating the feasibility of the technique and the analysis of the normal appearance of the cerebral white matter (CWM) in dogs in vivo. The analysis of the images in different planes and the three-dimensional reconstructions of the fiber tracts revealed a visual difference in the normal cerebral white matter appearance in one of the nine healthy dogs. This dog showed an altered topography of the corticospinal tract and corpus callosum due to displacement of the fibers. Quantitative assessment of ventricular volume in relation to the brain volume showed that this dog had larger lateral ventricles in relation to the other dogs. Authors believe this was most likely a normal variant because the dog exhibited no clinical neurological signs at the time of imaging. Other studies have shown that some canine species may exhibit a broad range of normal cerebral ventricular sizes, as assessed by neuroimaging, and that cerebral ventricular size is quite variable in a normal dog [1].

The only values quantified in the current study were those specific to the fiber tract and not to the ROIs because our research focused on demonstrating the feasibility of DTT for displaying and the normal appearance of the most representative cerebral white matter fiber tracts of healthy dogs in vivo. The quantification of FA and ADC values, or the exact description of the ROIs, were not used to visualize fiber tracts, as in our experience these values and descriptions can be obtained at different points of the fiber tract trajectories using the colored map. Future studies are needed to develop a method for preparing DTT templates and to reconstruct cerebral white matter pathways in the healthy dog in vivo using an ROI approach. This method is similar to the method used in humans and provides virtual representations of cerebral white matter tracts that are faithful to the classical ex vivo descriptions that include a detailed anatomical study of canines. The preparation of cerebral white matter templates for the healthy dog in vivo will allow investigators to follow the trajectory of the fibers delineating ROIs in DTI for DTT reconstruction. Improved knowledge of anatomy and the development of a template as a guide for the placement of ROIs could be used in future applications such as teaching and guiding virtual dissections of cerebral white matter tracts in healthy dogs in vivo, and comparison with pathological cases where the anatomy is observed distorted by the underlying disease process. Our findings are preliminary and future research will be needed to evaluate the use of DTT in clinical cases. Indeed, the use of this technique as a diagnostic tool is currently limited because templates and standardized FA and ADC values of the cerebral white matter of healthy dogs in vivo (which are prerequisites for the use of this technique in pathological cases) are lacking. Three previous canine DTI studies were conducted ex vivo, and one was conducted in vivo. In these studies, DTI was used to examine the structure and microstructure of the cerebral white matter but not for diagnostic purposes in canines. The findings of one study supported our study in that they also revealed the presence of association, commissural, and projection fibers in the dog. These data and information are also available in humans, and this allowed us to identify, compare, and reconstruct different fiber tracts in our healthy dogs in vivo. The results of study showed that the DTT reconstructions allowed identification and differentiation of CWM tracts in dogs in vivo, and that these reconstructions were comparable with those obtained in ex vivo canine studies. As new MRI options become increasingly available for daily clinical veterinary practice, novel diffusion techniques such as DTT warrant further exploration. In conclusion, findings from the current study indicated that DTT is a feasible noninvasive technique for in vivo study of CWM fiber tracts in the dog. We believe that the implementation of DTT as a noninvasive diagnostic method will complement conventional MRI, thus allowing investigators to examine the microstructural characteristics of the brain in vivo, and to obtain information on the anatomy, connectivity, and morphology of possible damage to fiber tracts in dogs suffering intracranial pathology or injury. Future research is needed to develop standardized ROI templates for in vivo canine studies and to compare DTT findings with confirmed pathologic findings [1].

Acknowledgements

We acknowledge Jose Agustin Moreno-Larios, Dr. Jaime Alonso Navarro Hernández, Dr. José Angel Gutiérrez-Pabello, and Dr. Jesús Taboada-Barajas for their advice and guidance during this work. Parts of this chapter are reproduced from Ref. [1].

Author details

Mitzi Sarahi Anaya García[1*] and Jael Sarahi Hernández Anaya[2]

*Address all correspondence to: hospitalveterinarioimagen@hotmail.com

1 Hospital Imagen, Distrito Federal, Mexico

2 College of Veterinary Medicine and Animal Science, National Autonomous University of Mexico (UNAM), Distrito Federal, Mexico

References

[1] Anaya García, M. S., Hernández Anaya, J. S., Marrufo Meléndez, O., Velázquez Ramírez, J. L., Palacios Aguiar, R. In Vivo study of cerebral white matter in the dog

using diffusion tensor tractography. Veterinary Radiology & Ultrasound. 2015;56(2): 188–195. doi:10.1111/vru.12211

[2] Pease, A., Miller, R. The use of diffusion tensor imaging to evaluate the spinal cord in normal and abnormal dogs. Veterinary Radiology & Ultrasound. 2011;52(5):492–497. doi:10.1111/j.1740-8261.2011.01837.x

[3] Alexander, A. L., Lee, J. E., Lazar, M., Field, A. S. Diffusion tensor imaging of the brain. Neurotherapeutics. 2007;4(3):316–329. doi:10.1016/j.nurt.2007.05.011

[4] Fujiyoshi, K., Konomi, T., Yamada, M., Hikishima, K., Tsuji, O., Komaki, Y., Okano, H. Diffusion tensor imaging and tractography of the spinal cord: from experimental studies to clinical application. Experimental Neurology. 2013;242:74–82. doi:10.1016/j.expneurol.2012.07.015

[5] Dauguet, J., Peled, S., Berezovskii, V., Delzescaux, T., Warfield, S. K., Born, R., Westin, C. F. Comparison of fiber tracts derived from in-vivo DTI tractography with 3D histological neural tract tracer reconstruction on a macaque brain. Neuroimage. 2007;37(2):530–538. doi:10.1016/j.neuroimage.2007.04.067

[6] Johansen-Berg, H., Behrens, T. E. Just pretty pictures? What diffusion tractography can add in clinical neuroscience. Current Opinion in Neurology. 2006;19(4):379. doi: 10.1097/01.wco.0000236618.82086.01

[7] Partridge, S. C., Mukherjee, P., Berman, J. I., Henry, R. G., Miller, S. P., Lu, Y. Vigneron, D. B. Tractography-based quantitation of diffusion tensor imaging parameters in white matter tracts of preterm newborns. Journal of Magnetic Resonance Imaging. 2005;22(4): 467–474. doi:10.1002/jmri.20410

[8] Takahashi, E., Dai, G., Rosen, G. D., Wang, R., Ohki, K., Folkerth, R. D., Grant, P. E. Developing neocortex organization and connectivity in cats revealed by direct correlation of diffusion tractography and histology. Cerebral Cortex. 2011;21(1):200–211. doi: 10.1093/cercor/bhq084

[9] Takahashi, E., Dai, G., Wang, R., Ohki, K., Rosen, G. D., Galaburda, A. M., Wedeen, V. J. Development of cerebral fiber pathways in cats revealed by diffusion spectrum imaging. Neuroimage. 2010;49(2):1231–1240. doi:10.1016/j.neuroimage.2009.09.002

[10] Ronen, I., Kim, K. H., Garwood, M., Ugurbil, K. Kim, D. S. Conventional DTI vs. slow and fast diffusion tensors in cat visual cortex. Magnetic Resonance in Medicine. 2003;49(5):785–790. doi:10.1002/mrm.10431

[11] Jacqmot, O., Van Thielen, B., Fierens, Y., Hammond, M., Willekens, I., Schuerbeek, P. V., Vanbinst, A. Diffusion tensor imaging of white matter tracts in the dog brain. The Anatomical Record. 2013;296(2):340–349. doi:10.1002/ar.22638

[12] Preti, M. G., Di Marzio, A., Mastropietro, A., Aquino, D., Baselli, G., Laganà, M. M., Spreafico, R. Tractographic reconstruction protocol optimization in the rat brain in-vivo: towards a normal atlas. In: 2011 Annual International Conference of the IEEE

Engineering in Medicine and Biology Society. 2011;8467–8470. doi:10.1109/IEMBS. 2011.6092089

[13] Griffanti, L., Baglio, F., Preti, M. G., Cecconi, P., Rovaris, M., Baselli, G., Laganà, M. M. Signal-to-noise ratio of diffusion weighted magnetic resonance imaging: Estimation methods and in vivo application to spinal cord. Biomedical Signal Processing and Control. 2012;7(3):285–294. doi:10.1016/j.bspc.2011.06.003

[14] Mishra, A., Lu, Y., Choe, A. S., et al. An image-processing toolset for diffusion tensor tractography. Magnetic Resonance Imaging. 2007;25:365–376. doi:10.1016/j.mri. 2006.10.006

[15] Mori, S., van Zijl, P. Fiber tracking: principles and strategies – a technical review. NMR in Biomedicine. 2002;15(7–8):468–480. doi:10.1002/nbm.781

[16] Beaulieu, C. The basis of anisotropic water diffusion in the nervous system - a technical review. NMR in Biomedicine. 2002;15(7–8):435–455. doi:10.1002/nbm.782

[17] Kim, J. H., Song, S. K. Diffusion tensor imaging of the mouse brainstem and cervical spinal cord. Nature protocols. 2013;8(2):409–417. doi:10.1016/j.jneumeth.2008.09.005

[18] Nimsky, C., Ganslandt, O., Merhof, D., Sorensen, A. G., Fahlbusch, R. Intraoperative visualization of the pyramidal tract by diffusion tensor imaging based fiber tracking. Neuroimage. 2006;30(4):1219–1229. doi:10.1016/j.neuroimage.2005.11.001

[19] Yoon, H., Park, N. W., Ha, Y. M., Kim, J., Moon, W. J., Eom, K. Diffusion tensor imaging of White and grey matter within the spinal cord of normal Beagle dogs: sub-regional differences of the various diffusion parameters. The Veterinary Journal. 2016;215:110–117. doi:10.1016/j.tvjl.2016.03.018

[20] Wang, P., Zhu, J. M. Quantitative diffusion tensor imaging of white matter microstructure in dog brain at 7 T. Open Medical Imaging Journal. 2010;4:1–5.

[21] Nucifora, P. G., Verma, R., Lee, S. K., Melhem, E. R. Diffusion-tensor MR imaging and tractography: exploring brain microstructure and connectivity. Radiology. 2007;245(2): 367–384. doi:10.1148/radiol.2452060445

[22] Hobert, M. K., Stein, V. M., Diziallas, P., Ludwing, D. C., Tipold, A. Evaluation of normal appearing spinal cord by diffusion tensor imaging, fiber tracking, fractional anisotropy, and apparent diffusion coefficient measurement in 13 dogs. Acta Veterinaria Scandinavica. 2013;55(1):36. doi:10.1186/1751-0147-55-36.

[23] Le Bihan, D., Breton, E., Lallemand, D., Grenier, P., Cabanis, E., Laval-Jeantet, M. MR imaging of intravoxel incoherent motions: application to diffusion and perfusion in neurologic disorders. Radiology. 1986;161(2):401–407. doi:10.1148/radiology. 161.2.3763909

[24] Asanuma, T., Doblas, S., Tesiram, Y. A., Saunders, D., Cranford, R., Pearson, J., Abbot, A., Smith, N., Towner, R. A. Diffusion tensor imaging and fiber tractography of C6 rat

glioma. Journal of Magnetic Resonance Imaging. 2008;28(3):566–573. doi:10.1002/jmri.21473

[25] Nickerson, J. P., Salmela, M. B., C. J., Andrews, T., Filippi, C. G. Diffusion tensor imaging of the pediatric optic nerve: Intrinsic and extrinsic pathology compared to normal controls. Journal of Magnetic Resonance Imaging. 2010;32(1):76–81. doi:10.1002/jmri.22228

[26] Glenn, O. A., Henry, R. G., Berman, J. I., Chang, P. C., Miller, S. P., Vigneron, D. B., Barkovich, A. J. DTI-based three-dimensional tractography detects differences in the pyramidal tracts of infants and children with congenital hemiparesis. Journal of Magnetic Resonance Imaging. 2003;18(6):641–648. doi:10.1002/jmri.10420

[27] Fujiyoshi, K., Konomi, T., Tsuji, O., Yamada, M., Hikishima, K., Momoshima, S. Okano, H., Toyama, Y., Nakamura, M. Assesment of injured spinal cord using diffusion tensor tractography. In: Uchida, K. editor. Neuroprotection and Regeneration of the Spinal Cord. Springer, Japan; 2014. pp. 345–366. doi:10.1007/978-4-431-54502-6

[28] Lori, N. F., Akbudak, E., Shimony, J. S., Cull, T. S., Snyder, A. Z., Guillory, R. K., Conturo, T. E. Diffusion tensor fiber tracking of human brain connectivity: aquisition methods, reliability analysis and biological results. NMR in Biomedicine. 2002;15(7–8):494–515. doi:10.1002/nbm.779

[29] Kleiser, R., Staempfli, P., Valavanis, A., Boesiger, P., Kollias, S. Impact of fMRI-guided advanced DTI fiber tracking techniques on their clinical applications in patients with brain tumors. Neuroradiology. 2010;52(1):37–46. doi:10.1007/s00234-009-0539-2

[30] Kerbler, G. M., Hamlin, A. S., Pannek, K., Kurniawan, N. D., Keller, M. D., Rose, S. E., Coulson, E. J. Diffusion-weighted magnetic resonance imaging detection of basal forebrain cholinergic degeneration in a mouse model. Neuroimage. 2013;66:133–141. doi:10.1016/j.neuroimage.2012.10.075

[31] Moldrich, R. X., Pannek, K., Hoch, R., Rubenstein, J. L., Kurniawan, N. D., Richards, L. J. Comparative mouse brain tractography of diffusion magnetic resonance imaging. Neuroimage. 2010;51(3):1027–1036. doi:10.1016/j.neuroimage.2010.03.035

[32] Catani, M., de Schotten, M. T. A diffusion tensor imaging tractography atlas for virtual in vivo dissections. Cortex. 2008;44(8):1105–1132. doi:10.1016/j.cortex.2008.05.004

[33] Lee, A. Y., Shin, D. G., Park, J. S., Hong, G. R., Chang, PH. H., Seo, J. P., Jang, S. H. Neural tracts injuries in patients with hypoxic ischemic brain injury: diffusion tensor imaging study. Neuroscience. 2012;528(1):16–21. doi:10.1016/j.neulet.2012.08.053.

[34] Pierce, T. T., Calabrese, E., White, L. E., Chen, S. D., Platt, S. R., Provenzale, J. M. Segmentation of the canine corpus callosum using diffusion tensor imaging tractography. AJR. American Journal of Roentgenology. 2014;202(1):W19. doi:10.2214/AJR.12.9791

[35] Konomi, T., Fujiyoshi, K., Hikishima, K., et al. Conditions for quantitive evaluation of injured spinal cord by in vivo diffusion tensor imaging and tractography: preclinical

longitudinal study in common marmosets. Neuroimage. 2012;63(4):1841–1853. doi: 10.1016/j.neuroimage.2012.08.040

[36] Cohen-Adad, J., Benali, H., Hoge, R. D., Rossignol, S. In vivo DTI of the healthy and injured cat spinal cord at high spatial and angular resolution. Neuroimage. 2008;40(2): 685–697. doi:10.1016/j.neuroimage.2007.11.031

[37] Cohen-Adad, J., Leblond, H., Delivet-Mongrain, H., Martinez, M., Benali, H., Rossignol, S. Wallerian degeneration after spinal cord lesions in cats detected with diffusion tensor imaging. Neuroimage. 2011;57(3):1068–1076. doi:10.1016/j.neuroimage.2011.04.068

[38] Gao, Y, Parvathaneni, P, Schilling, K. G., et al. A 3D high resolution ex vivo white matter atlas of the common squirrel monkey (Saimiri sciureus) based on diffusion tensor imaging. In: Styner, M. A., Angelini, E. D., editor. Medical Imaging 2016: Image Processing; March 21, 2016; San Diego California, United States. Spie Proceedings; 2016. Vol. 9784. doi:10.1117/12.2217325

[39] Gullapalli, J., Krejza, J., Schwartz, E. D. In vivo DTI evaluation of white matter tracts in rat spinal cord. Journal of Magnetic Resonance Imaging. 2006;24(1):231–234. doi: 10.1002/jmri.20622

[40] Ellingson, B. M., Kurpad, S. N., Li, S. J., Schmit, B. V. D. In vivo diffusion tensor imaging of the rat spinal cord at 9.4T. Journal of Magnetic Resonance Imaging. 2008;27(3):634–642. doi:10.1002/jmri.21249.

[41] Calebrese, E. Diffusion tractography in deep brain stimulation surgery: A review. Frontiers in Neuroanatomy. 2016;10:45. doi:10.3389/fnana.2016.00045